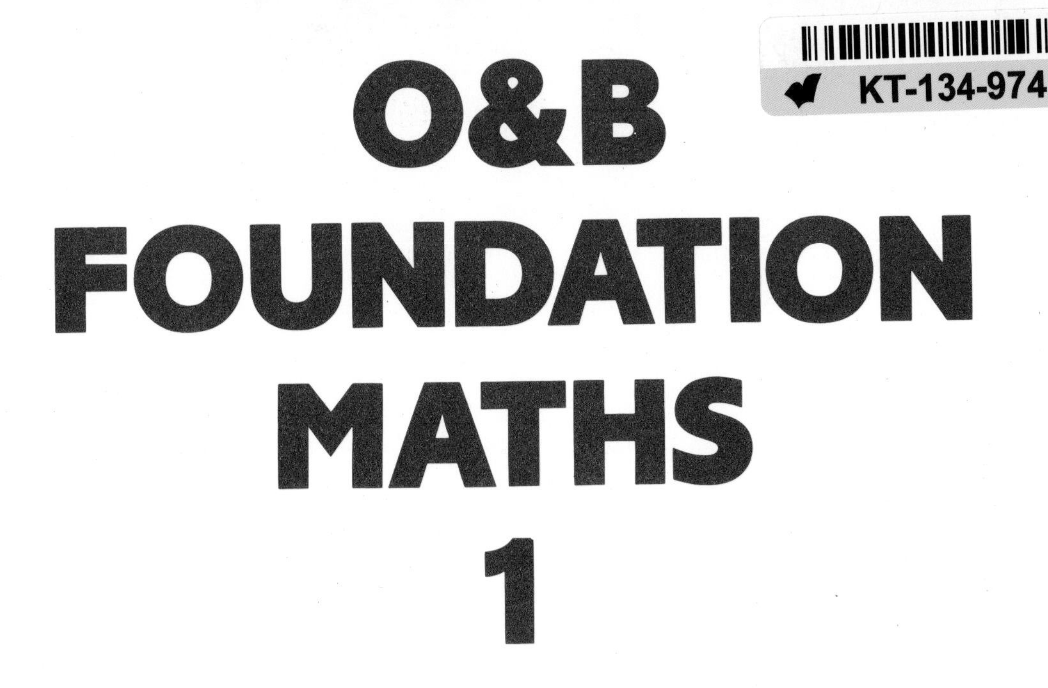

O&B FOUNDATION MATHS 1

JOHN B. BROWN

Oliver & Boyd

Illustrated by John Marshall, Donald Harley, Pamela Goodchild and Jake Sutton

Oliver and Boyd
Robert Stevenson House
1–3 Baxter's Place
Leith Walk
Edinburgh EH1 3BB

A Division of Longman Group Ltd

ISBN 0 05 003701 3

First published 1984

Set in 11/13 pt Univers Medium

Printed in Hong Kong by
Yu Luen Offset Printing Factory Ltd.,

Contents

1. TELEVISION

These pictures are from four different television programmes.
Do you recognise the programmes?

The four programmes shown on page 5 are:
Grandstand, BBC News, Dr Who and *Top of the Pops.*
These are all different **types** of programmes.
The table shows which type of programme each is.
There are many different types of programme.
Your teacher will discuss this with you.
Here is a page from a TV magazine to help you.

Programme	Type
Grandstand	Sport
BBC News	News
Dr Who	Serial
Top of the Pops	Music

12.30
The Sullivans
Adventures with the Australian Sullivan family and friends during World War Two.

1.0
News at One
News and views from around the world today.

1.30 A Plus
COOKING AT HOME
Secret of Good Soup
Today's recipes include garden vegetable soup, Stilton soup, and fresh watercress.

2.0 Racing from Newmarket
ITV's expert team brings you another sparkling card.

3.30
Blockbusters
Bob Holness with another round of the daily general knowledge quiz show.

4.0 Emu's World
Grotbags uses a secret weapon to capture Emu.

4.15 Doris
MARLON GOES CAMPING
Adventures with Doris, a thoroughly modern cat, and her boyfriend Marlon.

Jill Harvey (Jane Rossington) and Paul Ross (Sandor Elès) in Crossroads at 5.20.

4.20 On Safari
Join in the fun as contestants negotiate man-eating plants, tiger traps and swamps, and cross piranha-infested water.

4.45 Home
Drama revolving around the lives of children in care at an Australian community welfare home.

5.15 Job Spot
Produced in association with the Manpower Services Commission to guide viewers towards new jobs or new opportunities in training.

5.20 Crossroads
Paul Ross receives confidential information which involves him in an amorous situation.

News at 5.45
The latest national and international news from the studios of ITN in London with interviews and film reports, plus analyses of major events.

6.0
Scotland Today

6.30
Give Us a Clue
Host Michael Aspel and team captains Una Stubbs and Lionel Blair invite celebrities to play charades.

7.0 Winner Takes All
Jimmy Tarbuck hosts anot round of the general kno ledge gambling game.

7.30 Beyond the Poseidon Adventure

Capsized by freak wave, t giant passeng liner *Poseidon* floats upsi down in the Atlantic.

9.30 TV Eye
The week's big story – fro Alastair Burnet with Peter G Julian Manyon, Peter Prende gast and Denis Tuohy.

10.0 News at Ten

10.30 Studio
Jessica Langford demonstrate the arts and skills which g towards making animate films.

11.15 Late Call

11.20
The Mysteries of Edgar Wallace
INCIDENT AT MIDNIGHT

Dr Schroeder, a ex-surgeon now registered dru addict, is waiting in an all-nigh chemist's for a drug dosag when he recognises a ma called Leichner as Muller, a active Nazi in Vienna befor World War Two.

12.25 Closedown

ASSIGNMENT 1

Workbook page 1

Instead of listing the different types of programmes, we could draw a picture or a diagram to show this.

We could show the information in a diagram like the one below.
This has a centre with branches coming out, and is called a **branch** diagram.

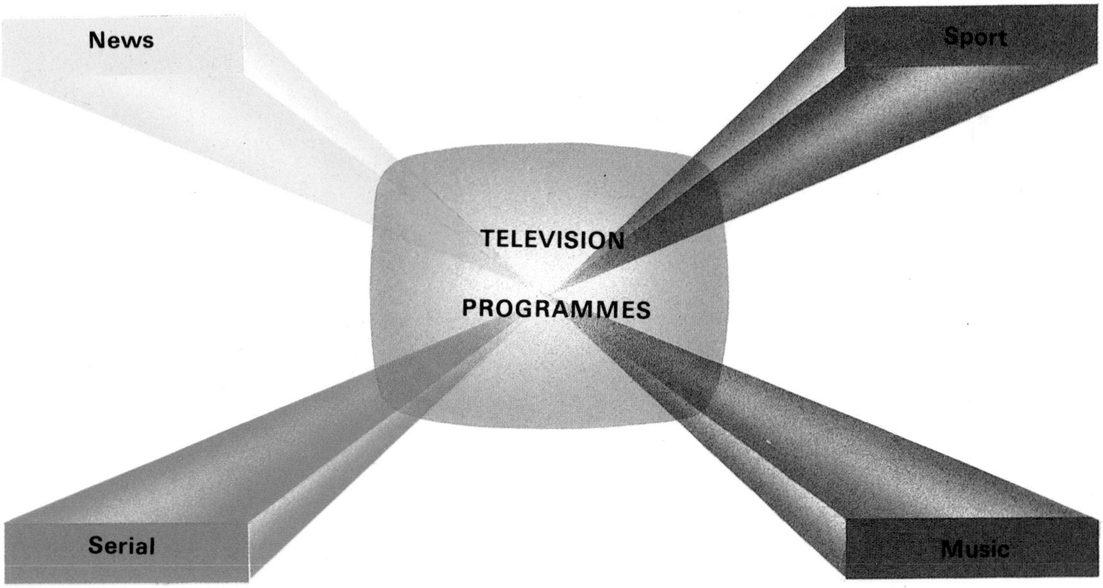

ASSIGNMENT 2

Workbook page 1

Finding Out More Facts

One way of finding out more facts is to ask people questions.
This is called **carrying out a survey**.
Let's carry out a survey to find which television programmes are the most popular.

Workbook page 2

When we find out the number of people who like a particular programme, we call this number the **frequency**.
If fifteen people like *Dallas*, we say the frequency is 15.

Workbook page 3

Drawing Graphs

A simple word for all the information you have collected is **data**.
It is possible to draw diagrams or **graphs** to illustrate this data.
Let's look at two types of graphs.

1 A **bar graph** shows the data in bars.
Let's assume someone collected the following information during a survey.

The information is in a table. The table shows the number of people who enjoyed the four television programmes. It is called a **frequency table**.

Programme	Frequency
Pot Black	12
It's a Knockout	8
Dallas	15
Dr Who	10

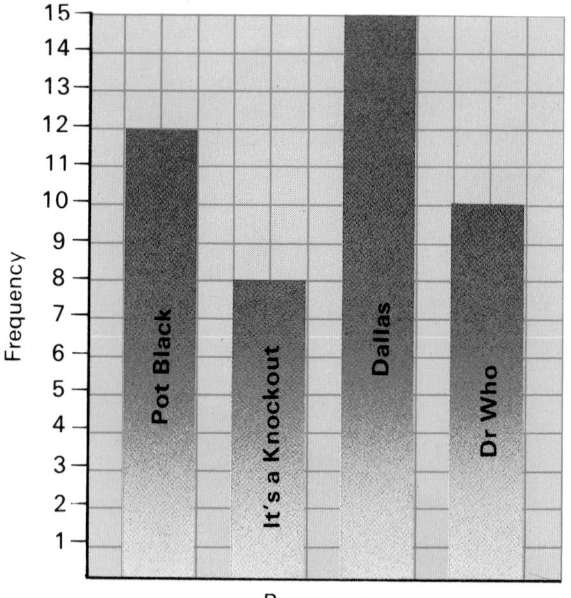

We can show this information in a bar graph.
A bar graph shows the frequency as bars.
Bar graphs are usually drawn on square paper.
The 'starting' lines are called **axes**.

ASSIGNMENT 5

Workbook page 4

2 A **block graph** shows the data in blocks. Let's assume someone else carried out a similar survey and collected the information shown in this frequency table.

Programme	Frequency
Top of the Pops	10
Crossroads	11
Tom and Jerry	7
Blankety Blank	9

This information is shown in the block graph opposite.

ASSIGNMENT 6

Workbook page 4

The information you collect does not have to be about TV programmes.
Here are some tables with information collected when different surveys were carried out.

1 *Survey:* To find the popularity of certain chocolate bars.

Data collected:

Chocolate Bar	Frequency
Mars	10
Galaxy	6
Milky Way	12
Topic	7

2 *Survey:* To find the popularity of certain drinks.

Data collected:

Drink	Frequency
Orange Juice	9
Lemonade	7
Tea	5
Coffee	6

3 *Survey:* To find favourite colours.

Data collected:

Colour	Frequency
Blue	5
Red	10
Yellow	8
Green	3

4 *Survey:* To find the most popular meal.

Data collected:

Meal	Frequency
Fish and Chips	10
Chicken	6
Steak and Kidney	6
Liver and Onions	2

5 *Survey:* To find the number of different types of vehicle on a road in half an hour.

Data collected:

Vehicle	Frequency
Car	11
Lorry	2
Bicycle	3
Tractor	1
Bus	4

6 *Survey:* To find out the most popular pet.

Data collected:

Pet	Frequency
Cat	8
Dog	12
Rabbit	3
Hamster	2
Goldfish	4

ASSIGNMENT 7

Workbook pages 5–6

You can use the data you collected earlier about TV programmes to find out more facts.
Get together with two or three of your classmates and combine the information you collected at home for Assignment 4.
Make up two tables; one showing the programmes selected by women, the other showing the programmes selected by men.
Draw two separate graphs to illustrate your data.

ASSIGNMENT 8

Workbook pages 6–7

Learning from the Data

When we look at the information we have collected, organised and illustrated, we can make some conclusions. For example, in the survey to find the popularity of certain types of chocolate bars, the following data was collected and you probably drew a bar graph similar to the one below.

Chocolate Bar	Frequency
Mars	10
Galaxy	6
Milky Way	12
Topic	7

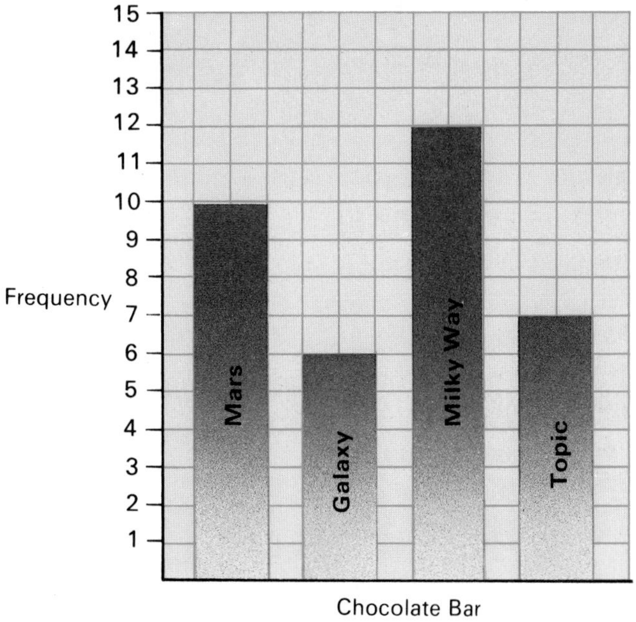

Your conclusions would be:
1 Milky Way is the most popular.
2 Galaxy is the least popular.
This is called **analysing the data**.

ASSIGNMENT 9

Workbook page 8

Where's the Maths?

You have:

1 collected information or data by carrying out a survey.
2 organised the information or data and counted the number of people who belonged to a group. The number of people is the frequency.
3 illustrated the information in bar graphs and block graphs.
4 used axes as your starting lines.
5 made comments about the results, or in other words analysed the data.

Statistics is a branch of maths where we:

1 collect
2 organise
3 illustrate
4 analyse

} information or data.

Words to Remember

survey	data	frequency	
bar graph	block graph	axes	analyse

2. SPORT

Do you know which sports are shown here?
Perhaps you recognise Daley Thompson and Martina Navratilova. You have probably seen them on TV.
Many different types of sports are shown on TV and are watched by millions.
Before you carry out a survey on sport, see how many different sports you can think of.

ASSIGNMENT 1

Workbook page 10

A Statistical Survey

You are now going to carry out a survey to find out some facts about people and sport. This time there are two questions.

1 Which sports do you enjoy watching?
2 Which sports do you enjoy playing?

Once you have collected all the information, you will be able to draw a bar graph and a block graph to illustrate it.

ASSIGNMENT 2

Workbook pages 11–12

Do you think adults enjoy the same sports as you? Find out by carrying out a survey among some adults you know.

ASSIGNMENT 3

Workbook page 13

Your class will now have a lot of information to organise.
Your teacher will help you do this. It will take quite a while!
When this is done, you can draw bar graphs to illustrate the data.

ASSIGNMENT 4

Workbook pages 14–15

On the Pitch and on the Court

Different sports have different sizes of pitches and courts.
Here are three diagrams of different pitches.

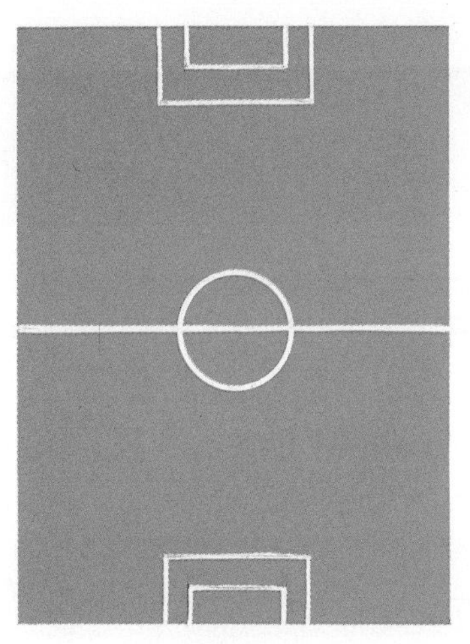

Football

Hockey

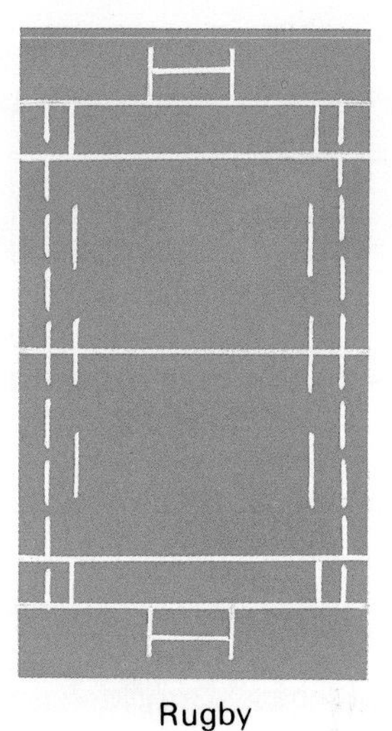

Rugby

Here are three diagrams of different courts.

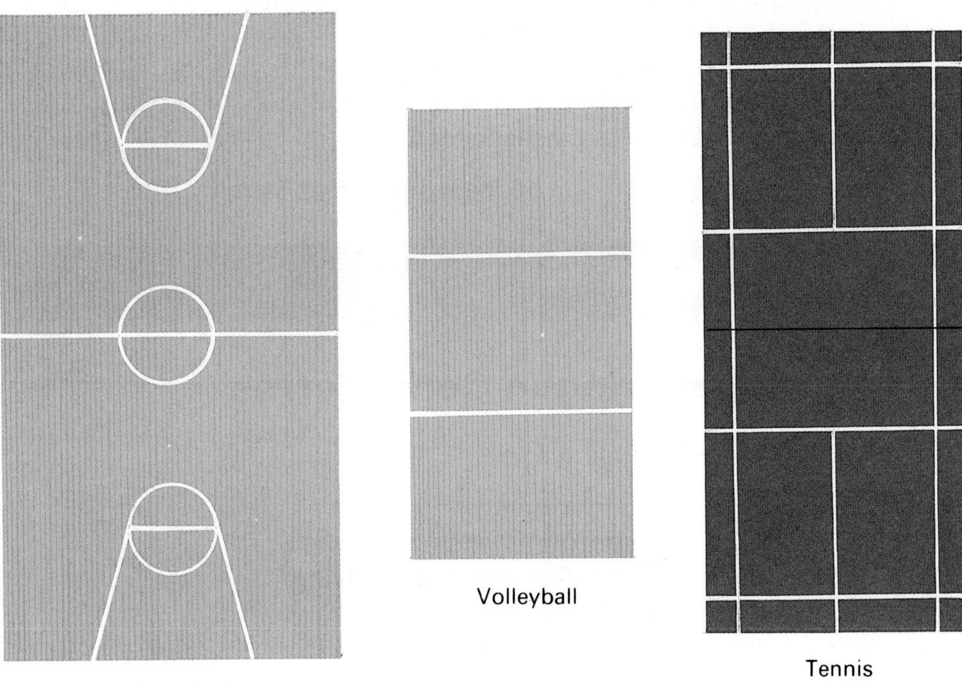

Basketball

Volleyball

Tennis

They are all **rectangles**. Most courts and pitches are **rectangular**.

For a groundsman to be able to mark out a pitch or court, he has to be able to **measure** the sides. To do this he would use a measuring tape. Before you use a measuring tape, let's look at how you measure a shorter distance.
You use a ruler.

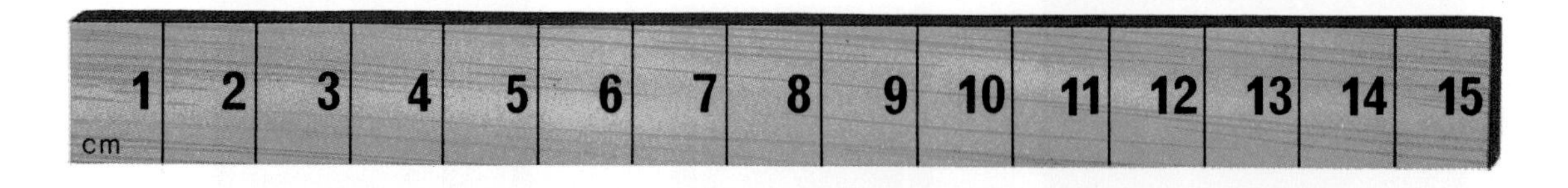

The units on the ruler are **centimetres**.

100 centimetres are equal to 1 **metre**.

To get an idea of these measurements you can make your own metre tape.

ASSIGNMENT 5

Workbook page 16

Instead of writing centimetre and metre in full, we often shorten the words:

centimetre becomes **cm**

metre becomes **m**

These are called **abbreviations**.
Abbreviations save time when writing measurements.

103 centimetres becomes 103 cm

54 metres becomes 54 m

ASSIGNMENT 6

Workbook page 17

Measuring in Centimetres

Measuring a straight line is quite easy using a ruler.

This line measures 5 cm.

This line measures 3 cm.

Here is a rectangle.
The longer side measures 5 cm, so we say the **length** is 5 cm.
The shorter side measures 3 cm, so we say the **breadth** is 3 cm.
These sizes are sometimes called the **dimensions** of the rectangle.

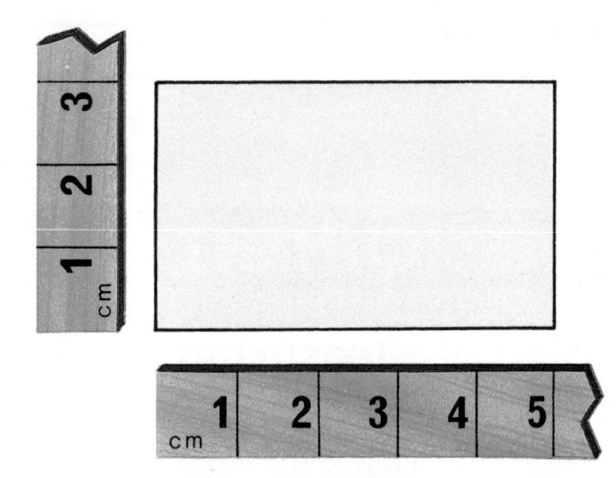

ASSIGNMENT 7 — Workbook page 17

You can use your metre tape to measure lines, but if you wanted to draw lines you would use a ruler. Rulers are usually made from perspex or wood.

ASSIGNMENT 8 — Workbook page 18

Being a Bit More Accurate

This line measures about 6 cm.
We can be more accurate than this if we want. Look at the ruler below.

Each centimetre is divided into ten smaller units.
Each of these smaller units is called a **millimetre**. The abbreviation for millimetre is **mm**.

So | 10 millimetres = 1 centimetre | or | 10 mm = 1 cm |

This line measures 5 cm 2 mm.
We can write this as 5.2 cm.

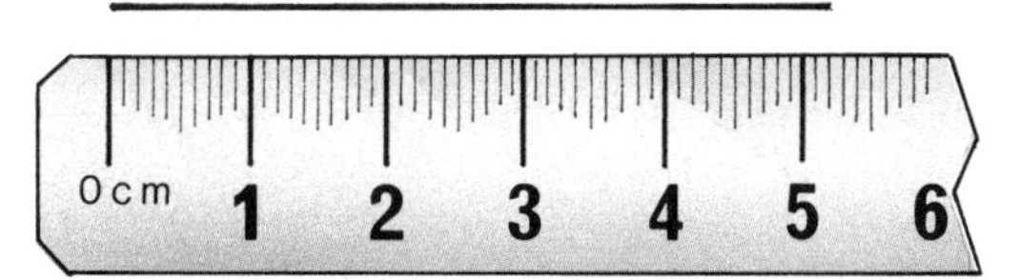

This line measures 6 cm 4 mm.
We can write this as 6.4 cm.

ASSIGNMENT 9

Workbook page 19

Measuring Pitches and Courts

Pitches and courts are measured in metres.
Look at the diagrams of the pitches and courts again.

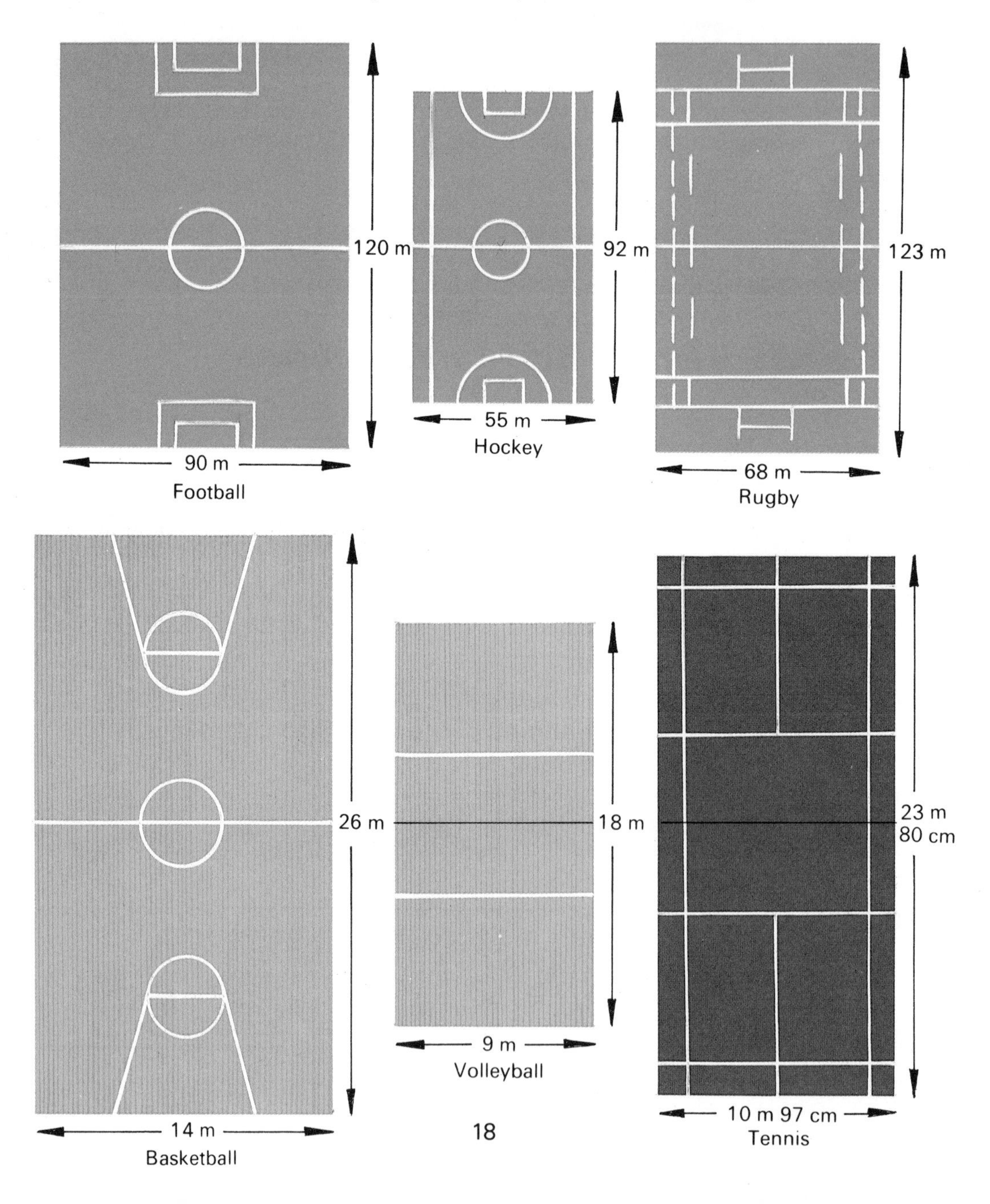

From these diagrams we can read off the dimensions of each pitch or court.
For example, the length of the football pitch is 120 m and its breadth is 90 m.
Some pitches and courts are not measured in exact metres. Look at the tennis court. Its length is 23 m 80 cm. Remember one metre is made up of 100 centimetres, so 23 m 80 cm can be written as 23.80 m. Similarly, 5 m 34 cm can be written as 5.34 m and 16 m 12 cm can be written as 16.12 m.

ASSIGNMENT 10

Workbook pages 19–20

The gym or sports hall in your school will be used for several different sports, each marked by coloured paint or tape. Can you identify each sport by the markings?

ASSIGNMENT 11

Workbook pages 20–21

Using the Right Unit

We use different units to measure different things.
We would measure the height of a door in metres.
We would measure the length of a pencil in centimetres.
We would measure the length of an insect in millimetres.

ASSIGNMENT 12

Workbook page 21

Where's the Maths?

You have:

1. carried out a survey and collected statistics.
2. illustrated the data in a graph.
3. analysed the data.
4. looked at the measurement of figures.
5. revised your units of measurement.
6. made a metre tape.
7. measured and drawn lines and rectangles using millimetres, centimetres and metres.
8. taken measurements from figures.

Words to Remember

survey statistics graph axes

measurement rectangle length breadth

metre (m) centimetre (cm) millimetre (mm)

Measurements to Remember

10 mm = 1 cm

100 cm = 1 m

3. MUSIC

Everybody listens to music.

Some people like pop music.

Some people like classical music.

Others prefer country and western.

There are many different types of music. Your teacher will let you listen to some of these now.

ASSIGNMENT 1 — Workbook page 23

You have listened to several different types of music. There are many more. Show some of these different types of music in a branch diagram.

A Statistical Survey

Now let's find out something about the musical interests of your class. We can do this by carrying out a survey to find:

1 the most popular singer; 2 the most popular group.

Sometimes when we draw bar graphs, it is not possible to write on the bars. Instead we write on the bottom axis. Here are two examples showing popularity ratings of groups and singers. The data is from the 1960s! Notice that there is no space between the bars this time.

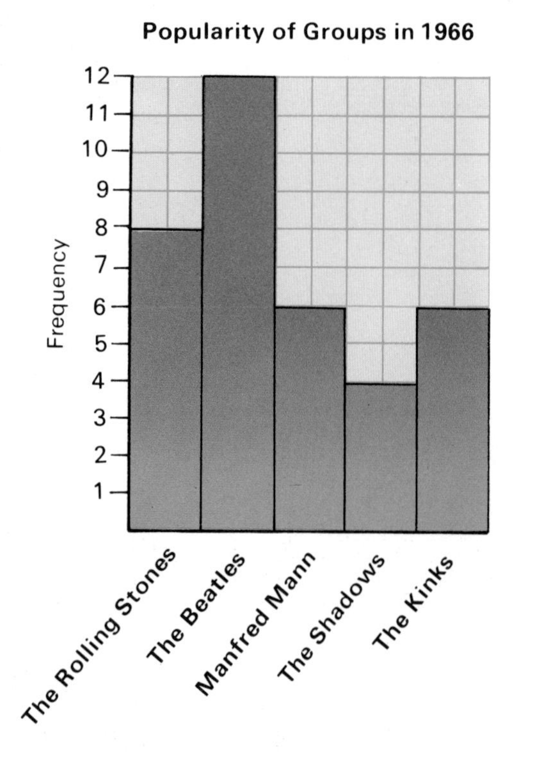

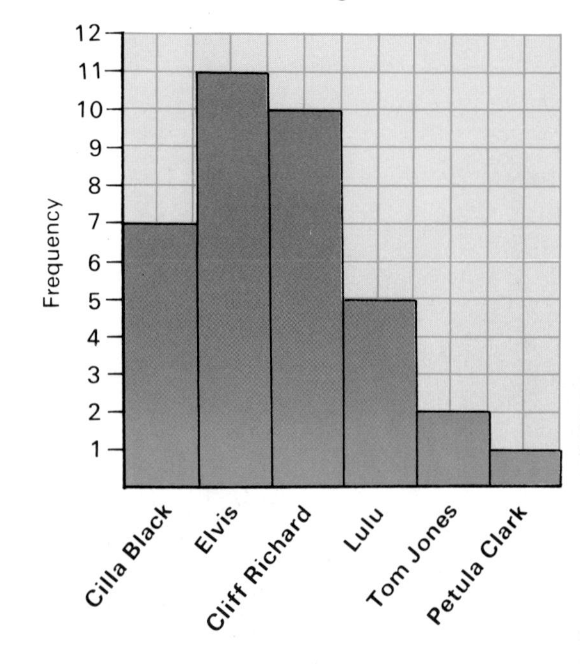

ASSIGNMENT 4

Workbook page 25

Now that you have drawn the bar graphs, you will see at a glance the most popular group and the most popular singer. What about your parents' generation? Do they have different favourites? Carry out a survey to find out.

ASSIGNMENT 5

Workbook page 26

To find out who the adults chose, your teacher will collect the information from all the survey sheets. You'll probably find the results are different from those of your class!

ASSIGNMENT 6

Workbook page 27

Rotations – Revolutions – Circles

All types of music are reproduced on tapes and records.

The first machines for recording and reproducing music were called phonographs. They did not play flat records. They had cylinders which turned. You can see these inventions in museums all over the country.

Modern sound systems are a lot more sophisticated! On your record player, stereo or music centre you have a turntable. It is **circular**. It turns or **rotates**.

On most modern turntables there are two speeds. One setting is used for long-playing records (LPs). The other setting is used for singles.
Early record players had usually only one speed. The turntable moved round much faster.
Record players in the 1950s and 1960s had three settings. Here are the three settings.

Type of Record	Speed
LPs	33
Singles	45
Old singles (called '78's)	78

LPs turn 33 times in one minute.
This is sometimes written 33 rpm.

What do the numbers mean?
The number tells us how often the record turns round in one minute.

Singles turn 45 times in one minute.
This is written 45 rpm.

rpm means revolutions per minute.
One revolution is the same as one complete turn.

ASSIGNMENT 7

Workbook page 28

Drawing Circles

Here is a picture of some different types of records. It is difficult to draw a circle freehand. Try to copy this picture and you'll find this out for yourself!

ASSIGNMENT 8

Workbook page 28

We can make it easier by using any of the following.

1 A coin or a tin lid.

All you need to do is draw round the circular shape. It's simple!

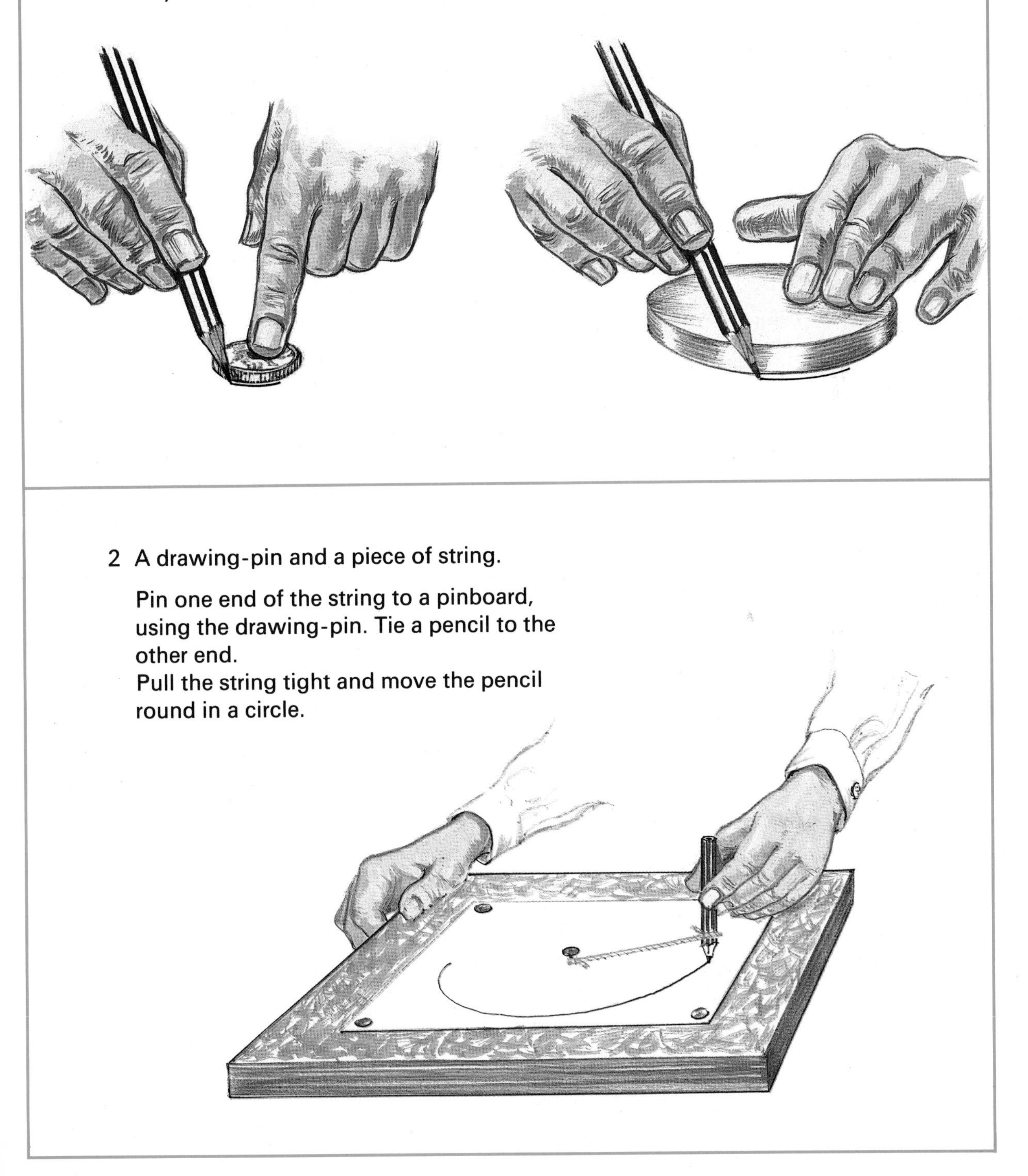

2 A drawing-pin and a piece of string.

Pin one end of the string to a pinboard, using the drawing-pin. Tie a pencil to the other end.
Pull the string tight and move the pencil round in a circle.

3 A pair of compasses.
This is the mathematical instrument for drawing circles. You have already made a metre tape. It is also possible to make your own compasses.

ASSIGNMENT 9

Workbook pages 29–31

The Size of a Circle

You can see that Circle *A* is smaller than Circle *B*. But how can you measure the exact size of a circle? You have already measured the length and breadth of a rectangle, but a circle does not have a length or breadth.

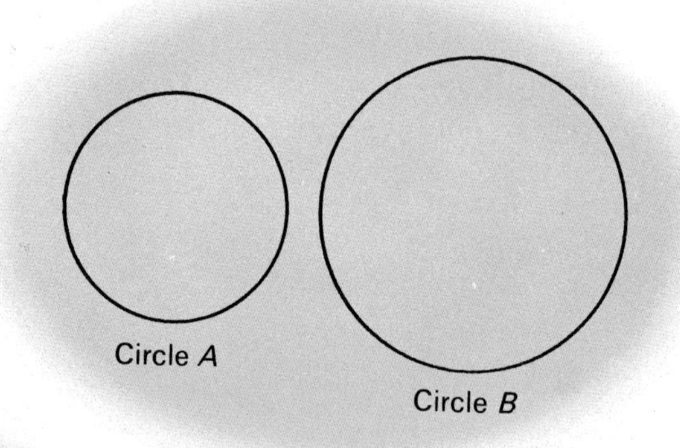

Look at this circle.

The point in the middle of the circle is called the **centre**.

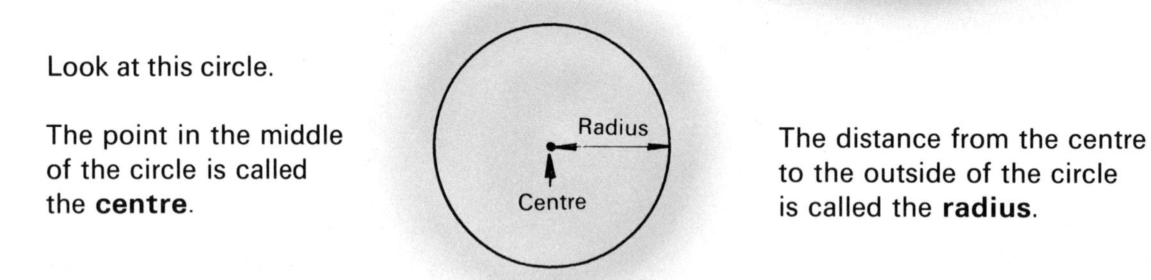

The distance from the centre to the outside of the circle is called the **radius**.

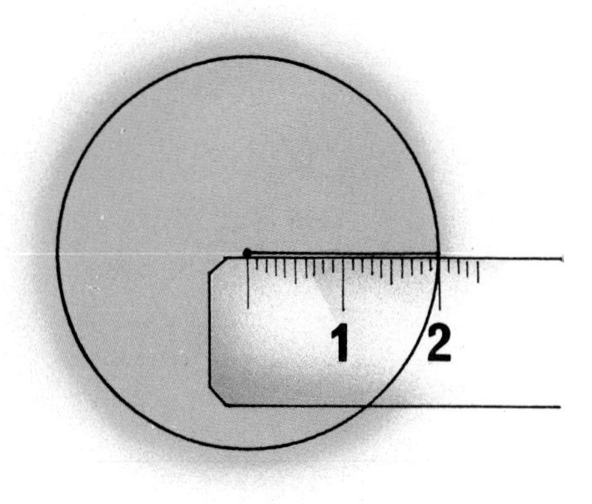

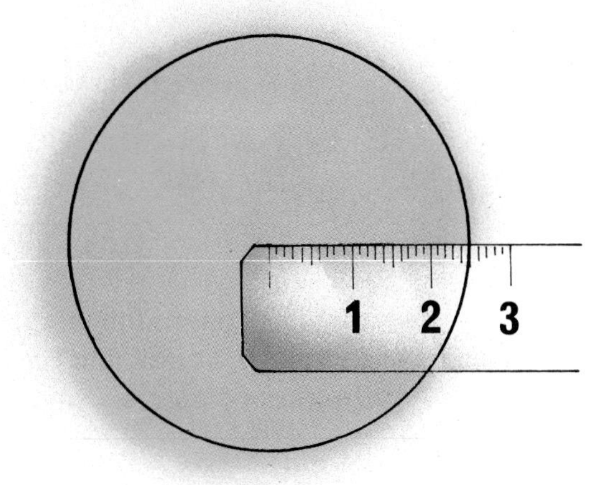

We can measure the radius using a ruler. The radius of this circle is 2 cm.

If the radius is not drawn in, all we need to do is measure from the centre to the outside of the circle. The radius of this circle is 2.5 cm.

ASSIGNMENT 10

Workbook page 32

You have already drawn circles using different 'instruments'.
Compasses are used to draw circles when we are given the radius.
Here is how to draw a circle with a radius of 4 cm.

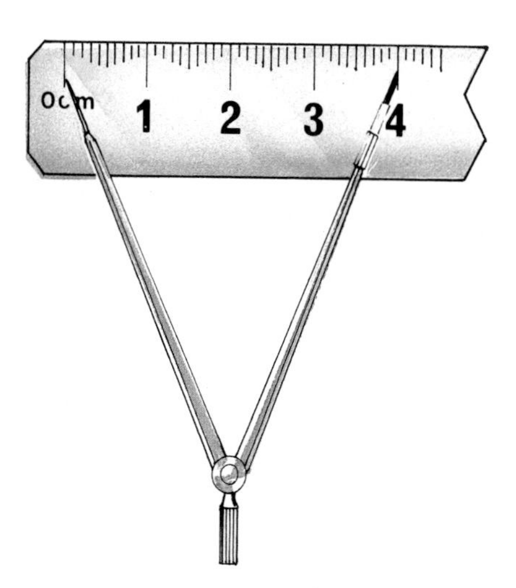

1 Place the point of the compasses at 0 cm on the ruler.
2 Pull the compasses apart until the pencil is at 4 cm on the ruler. Make sure the point is still at 0 cm.
3 Now place the point on the paper and draw the circle.

ASSIGNMENT 11

Workbook page 33

Finding the Centre

If you were asked to find the length of the radius of this circle, you would have a problem. There is no centre to measure from.

Here is one way to find the centre of a circle.

1 Cut out the circle and carefully fold it in two. Make sure there is no overlap at the edge.

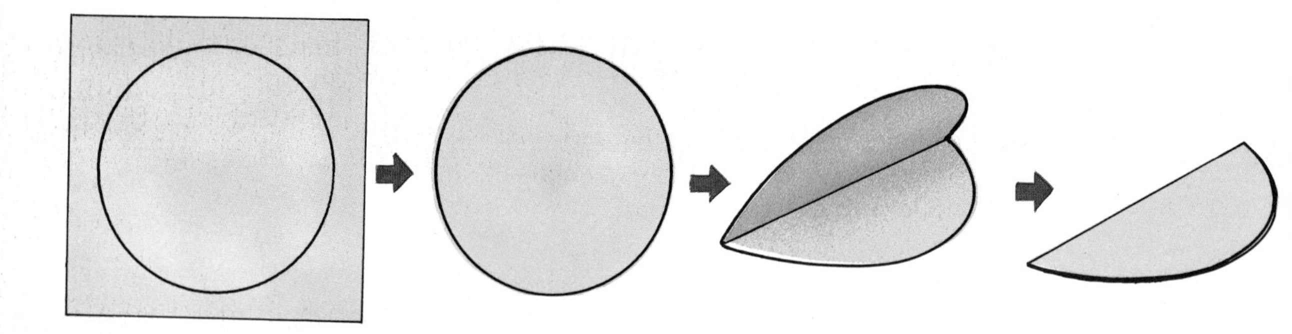

2 Open the circle out. You now have a circle with a fold line down the middle.

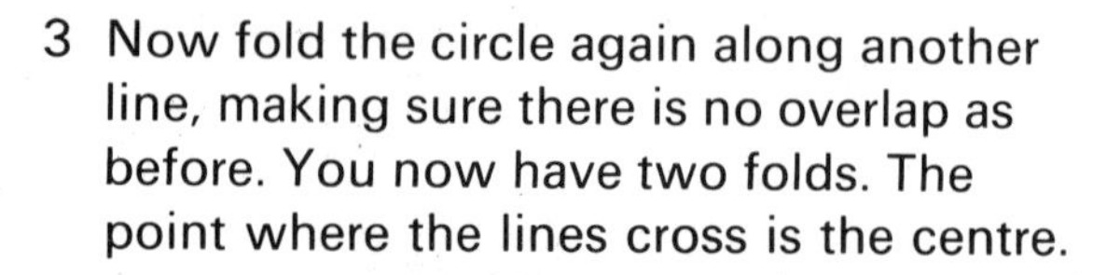

3 Now fold the circle again along another line, making sure there is no overlap as before. You now have two folds. The point where the lines cross is the centre.

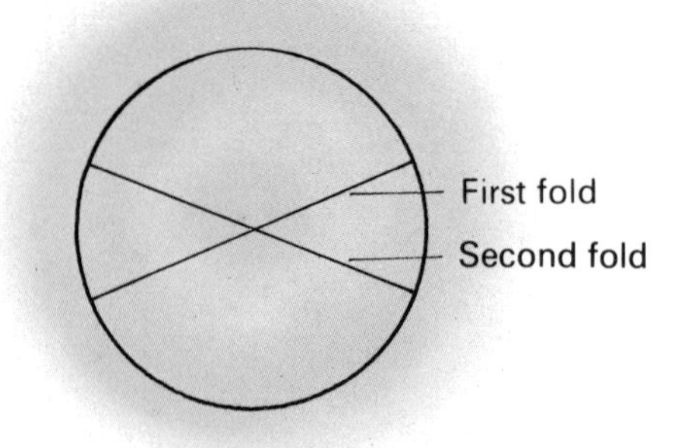

ASSIGNMENT 12

Workbook page 34

Round and Rotating Objects

This picture shows the rings round the planet Saturn

There are many objects which are round and which rotate. You know about records and tapes. See how many others you can think of. Try to find examples by looking in some magazines and newspapers.

ASSIGNMENT 13

Workbook page 34

Where's the Maths?

You have:

1 carried out a survey and collected statistics.
2 illustrated the data in a graph.
3 analysed the data.
4 looked at turntable speeds on a record player and understood what they mean.
5 learned that 33 rpm means 33 revolutions per minute.
6 drawn circles using compasses.
7 measured the radius of a circle.

Words to Remember

rotation **revolution** **circle**

circular **radius** **compasses**

4. THE FAMILY

Here is the Brown family. Margaret and Bill Brown have three children called Emma, Kirsty and Stewart.

Bill's mum and dad are called Lena and Eric Brown.

Margaret's mum and dad are called Jean and John Paterson.

We can draw a diagram to show how all the Browns and Patersons are related.

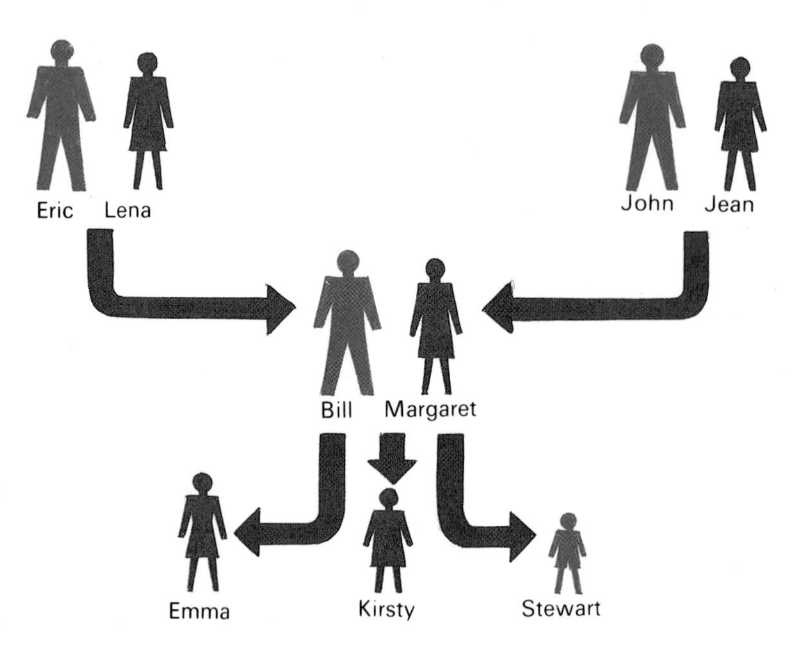

We can make this even simpler.

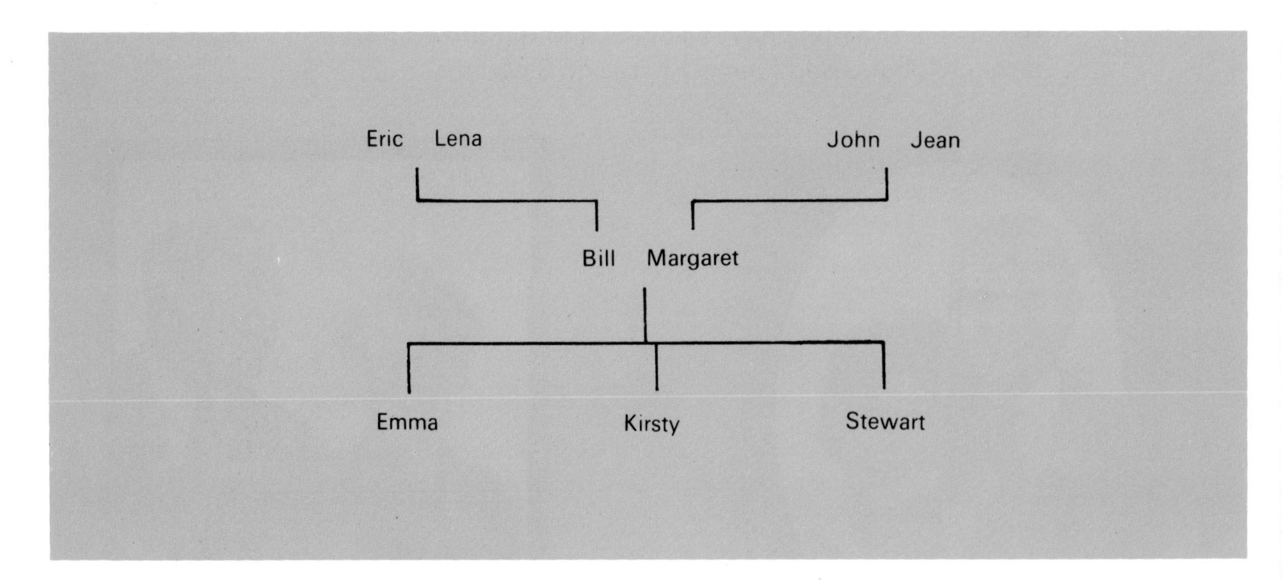

This is called a **family tree**. But a family tree usually has more than just names. Dates are written beside the names.

1 When the person was born.

2 When the person died.

For example, 1930–1982 means that a person was born in 1930 and died in 1982.

Read the following list of events in the life of the Muir family.

1901 George Muir was born.
1920 George married Mary Hepburn (born 1902).
1925 They had a son, Peter.
1930 They had another son, Andrew.
1948 Peter married Angela Watson (born 1928).
They had two children.
1949 Jennifer was born.
1950 James was born.
1950 Andrew married Isobel Boyter (born 1931).
They had three children.
1952 Tom was born.
1954 Kathleen was born.
1957 Amanda was born.
1969 George Muir died.
1980 Mary Muir died.

There are a lot of names and a lot of dates here! We can draw a family tree to illustrate the family history.

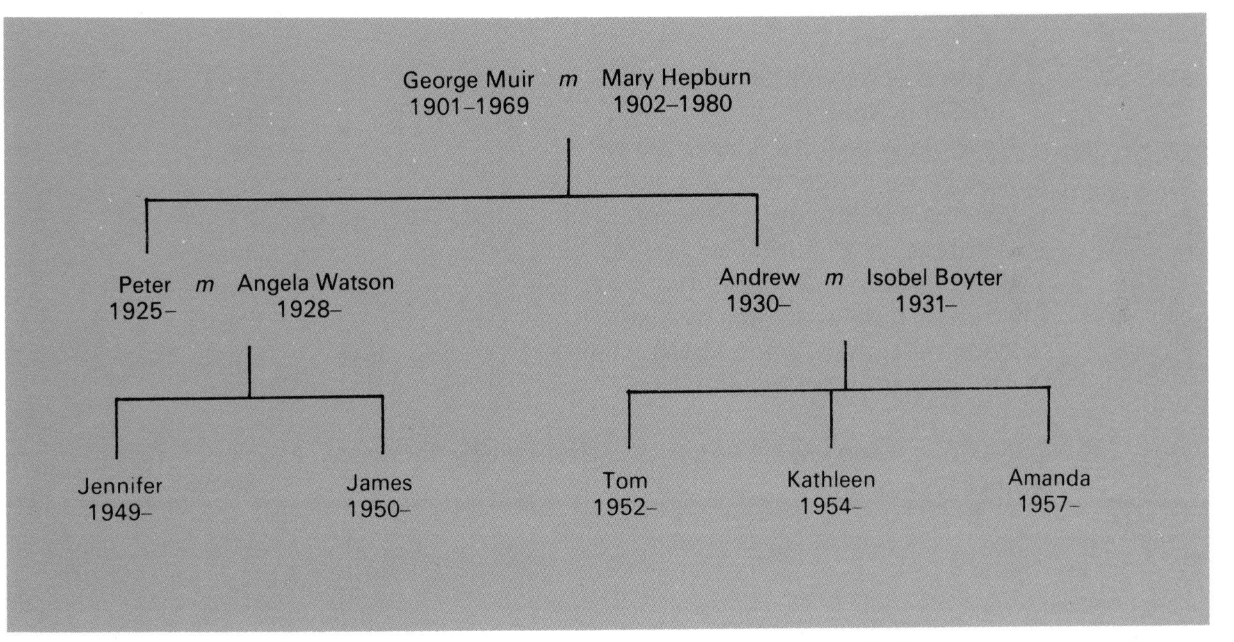

Note: 'm' means married. Marriage dates are not recorded.

Family trees are time diagrams. They tell us when family events took place. They also tell us how members of a family are related.

ASSIGNMENT 1

Workbook page 36

Taking Information from a Family Tree

This is part of the Smith family tree.

John Smith 1924–1979 *m* Sheila Baxter 1926–

Simon 1945–

Samuel 1946– *m* Lesley Hudson 1942–

Michael 1970–

Alan 1975–

Sandra 1949– *m* David Hynd 1952–

Sarah 1972–

Here are some questions with answers taken from the tree. Cover up the answers and then look to see if your own answers are correct.

1	When was Samuel born?	1946
2	Did Simon marry?	No
3	How old was John when he died?	55
4	Was Samuel older or younger than Lesley?	Younger
5	What is the difference between Michael and Alan's ages?	5 years
6	What is Sarah's surname?	Hynd
7	What relation is Michael to John?	Grandson
8	What relation is Samuel to Alan?	Father
9	What relation is Simon to Michael?	Uncle
10	What relation is Sheila to John?	Wife

If you got some answers wrong and don't understand why, ask your teacher to explain.

ASSIGNMENT 2

Workbook pages 37–38

Your Own Family Tree

Anyone can draw their own family tree. All you have to do is carry out a survey! You will have to find out the names, dates of birth and dates of any deaths in your family.

Workbook pages 38–39

It is easier to follow up one side of a family – either your father's side or your mother's side. If you do this you can go even further back in the history of your family.
Try to find out some information about your family going back further than your grandparents. Your teacher will help you construct your family tree, once you have gathered the information.

ASSIGNMENT 4

Workbook page 39

Dates

In your family tree you will have used dates such as 1939, 1950 and 1982. The year 1939 (nineteen thirty-nine) is one thousand, nine hundred and thirty-nine years after the birth of Christ. When we write a date which is after his birth, we sometimes put AD before the year.

In AD 43 the Romans invaded Britain.
AD is the abbreviation for *Anno Domini*.
This is latin for 'In the year of our Lord'.

If we want to talk about a date before Christ was born, we put BC after the year.

In 776 BC the first Olympic Games were held in Greece.
BC is the abbreviation for 'Before Christ'.

ASSIGNMENT 5

Workbook page 40

Here are some important events in our history. The dates tell us when they took place. The events all took place after Christ was born. Usually we don't have to put AD in front of them.

Battle of Bannockburn
1314

Battle of Britain
1940

Jacobite Rebellion
1745

Massacre of Glencoe
1692

Spanish Armada
1588

Battle of Trafalgar
1805

Battle of Waterloo
1815

Magna Carta
1215

First Crusade to Holy Land
1096

Battle of Agincourt
1415

Falklands War
1982

Battle of Flodden
1513

These events took place at different times in our history. Often it is helpful to put dates like these in a **chronological** order. To do this we put the oldest event first and the most recent event last. Here are some more historical events. This time they are in chronological order.

1605 Guy Fawkes executed
1666 The Great Fire of London
1789 The French Revolution started
1807 The Slave Trade abolished
1914 The First World War started
1953 Queen Elizabeth II crowned

ASSIGNMENT 6

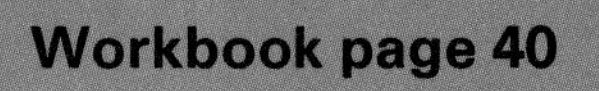
Workbook page 40

Measuring Time

We measure time in many different units.
We have been talking about years. A year is how long it takes the earth to move round the sun.

Earth
Sun

Here are some time spans usually measured in years.

1 How long you will live.
2 How long you stay at school.
3 How long it takes to build a motorway.
4 How long it takes for a seedling to grow into a tree.

In history we often measure time in centuries.

One century = 100 years.

We are now living in the twentieth century. The first century was from the birth of Christ to the beginning of the year 100. Here is a list of some of the centuries so far.

AD 100→AD 199 second century
AD 500→AD 599 sixth century
AD 1200→AD 1299 thirteenth century
AD 1800→AD 1899 nineteenth century
AD 1900→AD 1999 twentieth century

In 1770 Captain Cook landed in Botany Bay, Australia. This took place in the eighteenth century.

In 1897 Queen Victoria celebrated her Diamond Jubilee. This took place in the nineteenth century.

In 1969 man first landed on the moon. This took place in the twentieth century.

ASSIGNMENT 7

Workbook page 41

Life Spans

Here are five famous Scots.
The information tells us: 1 when they lived (their life spans);
2 what they invented or discovered;
3 the year of their invention or discovery.

From this we can work out how long each of them lived, and what age they were when they made their invention or discovery.
For example, John Logie Baird was born in 1888 and died in 1946.
Look at this 'time-line'.

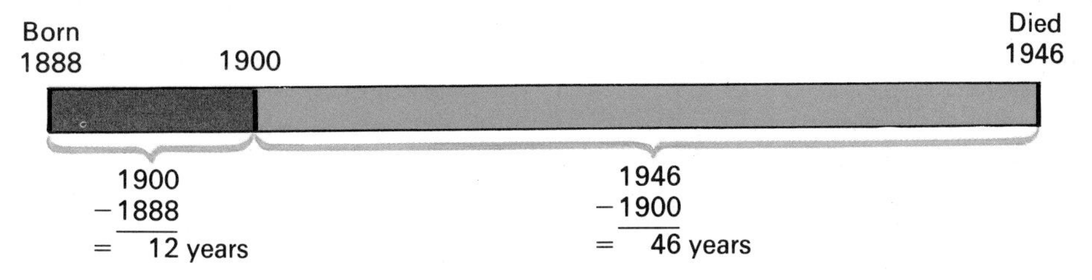

Total = 12 + 46 = 58 years. So he was 58 years old when he died.
The first television was demonstrated in 1926. 1888 to 1926 is 12 + 26 = 38 years. So he was 38 years old in 1926.

ASSIGNMENT 8

Workbook page 42

Other Time Spans

We can write time spans for different events.
The First World War lasted from 1914–18. It lasted for four years.
Notice that as this time span starts and finishes in the same century, we do not need to repeat the '19'.
Sir Francis Drake sailed round the world from 1577–80. It took him three years to sail round the world.

Here are some kings and queens. Their years of rule are given below.

William (The Conqueror)
1066–87

Henry VIII
1509–47

Elizabeth I
1558–1603

James I (VI of Scotland)
1566–1625

Charles II
1660–85

George III
1760–1820

Victoria
1837–1901

Edward VII
1901–10

To work out how long each one reigned, we subtract the second date from the first.

$$\begin{array}{r} 1087 \\ -1066 \\ \hline = \quad 21 \end{array} \text{ years}$$

William The Conqueror reigned for 21 years.

It is more difficult to work this out if their reign spans two centuries. Using a time-line makes it easier.

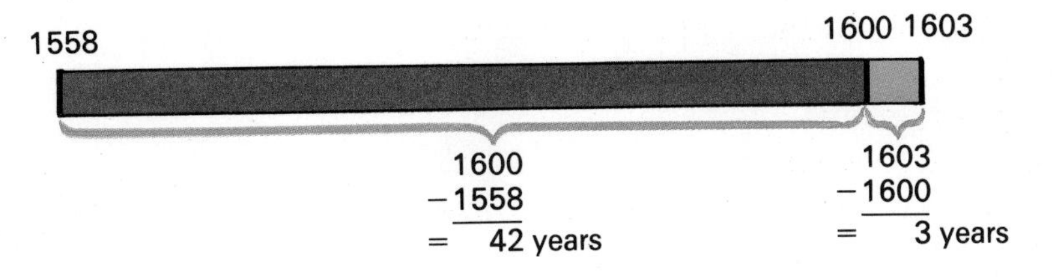

Elizabeth I reigned for $42+3=45$ years.

ASSIGNMENT 9

Workbook page 42

Now you know some facts about your family history. You should have picked up a few facts about Britain's history, too!

Where's the Maths?

You have:

1 Carried out a survey.
2 illustrated the data in a time diagram (family tree).
3 used the family tree to work out relationships.
4 looked at a time scale in history (BC and AD).
5 learned what one year is.
6 learned that 100 years = one century.
7 put dates in chronological order.
8 put events in the correct century.
9 used time spans to work out certain facts.

Words to Remember

family tree **AD** **BC** **year**

century **time span** **chronological**

5. THE HOME

Most families live in a house. This might be in a town, or in the countryside, or even by the sea. Where would you prefer to live?
There are many different types of houses. Some of these are shown on this page.

ASSIGNMENT 1

Workbook page 44

Not everybody lives in a house. Here are some pictures of other types of dwellings. Have you ever lived in any of these?

ASSIGNMENT 2

Workbook page 44

There are many other types of dwellings in which people live. Can you think of any?

ASSIGNMENT 3

Workbook page 44

A Statistical Survey

Let's carry out a survey to find out which type of house each person in your class lives in. Perhaps you all live in the same type of house!

ASSIGNMENT 4

Workbook page 45

Building Homes

This photograph was taken around the beginning of this century.
It shows houses on St Kilda, the most westerly island off Scotland.
Look at the materials used to build the houses.

Nowadays, most houses are built using materials such as bricks, mortar and concrete.

Before a house is built, drawings and diagrams are made. The person who designs the house and makes the drawings is called an **architect**. Drawings can be made showing the front, side and top of the house.

The most useful drawing is often the one looking down on top of the house with the roof removed! This is called the **plan**. Inside we can see the **layout** of the house. The plan is used by all the tradespeople (bricklayers, joiners, electricians, etc.) who help to build the house.

Here is the plan of a two-bedroomed house. Notice how the doors and windows are drawn.

This is a doorway:

This is a window:

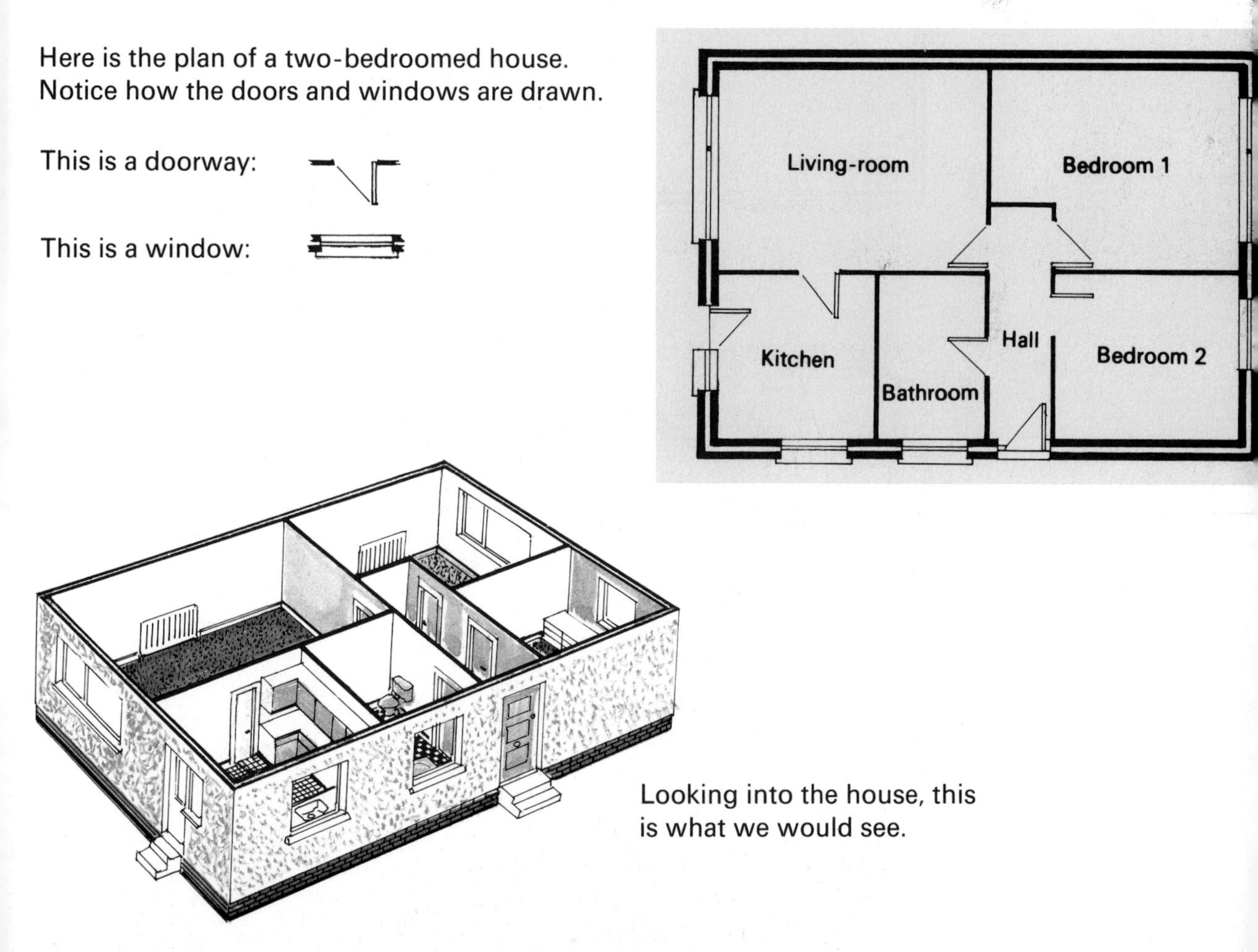

Looking into the house, this is what we would see.

Most houses are now constructed so that they are easy to look after and keep warm. To do this an architect must fit all the rooms into a certain space.

Here is a plan showing the outside walls of a house.
The windows and doors are shown.
There are five windows and two doors.

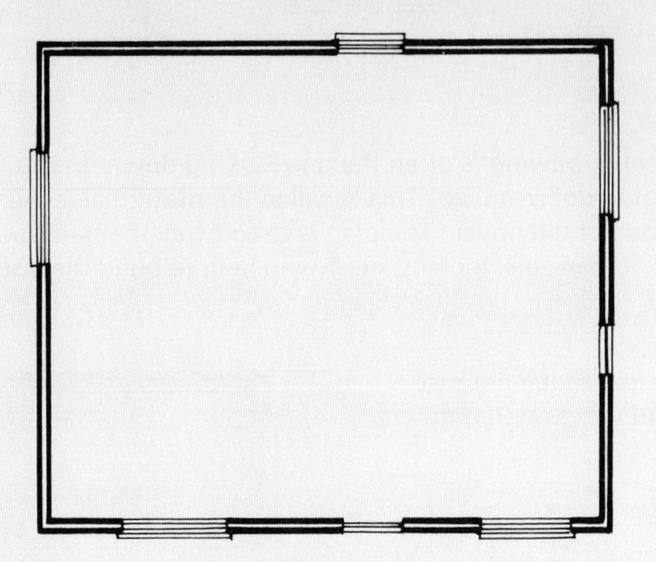

Here are the plans of the rooms that fit inside the house.

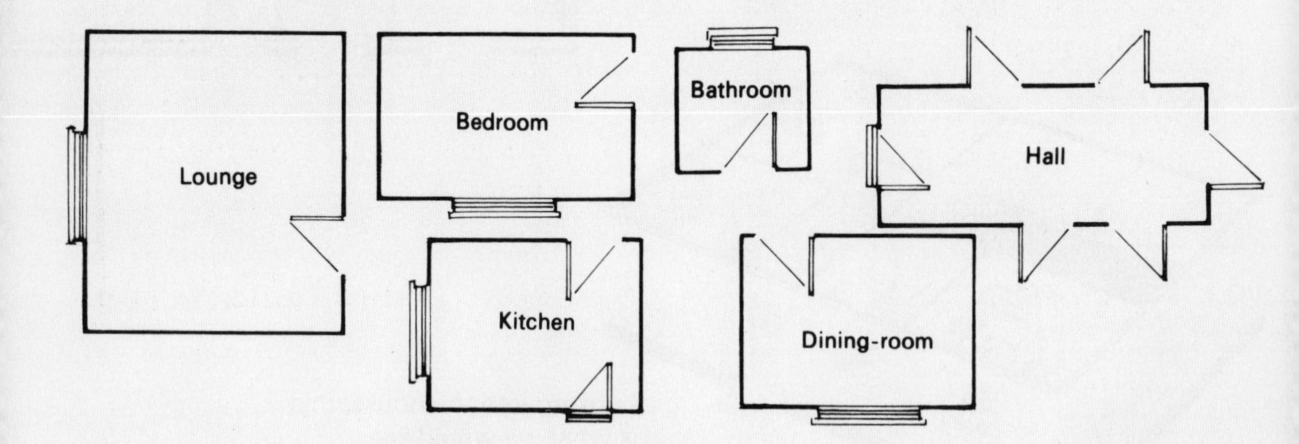

When we fit these into the space given, we end up with this:

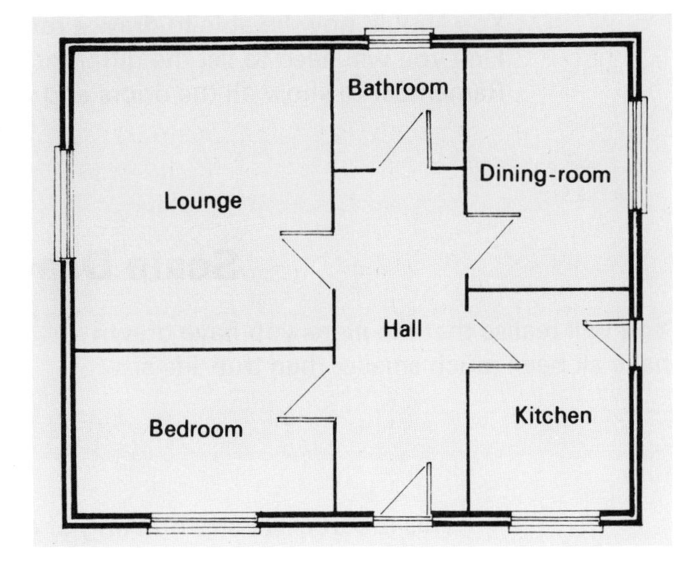

How many doors are there now?

ASSIGNMENT 6 Workbook pages 46–47

Drawing House Plans

Look at this diagram of the inside of a house. We could draw a rough plan of the house from this.

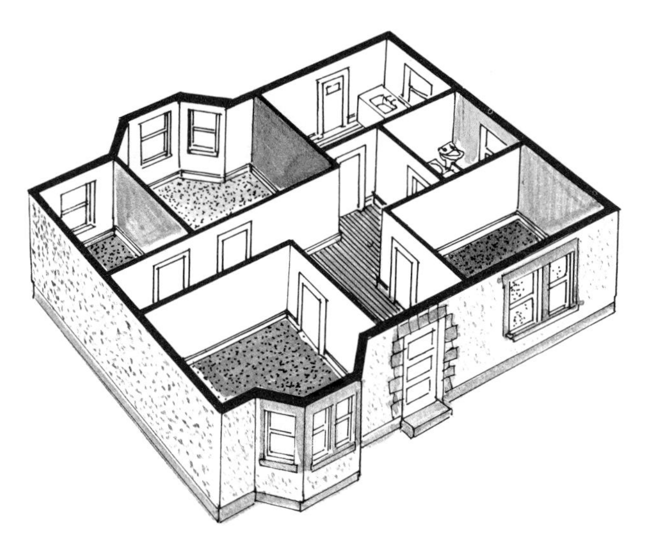

Here is what it would look like.

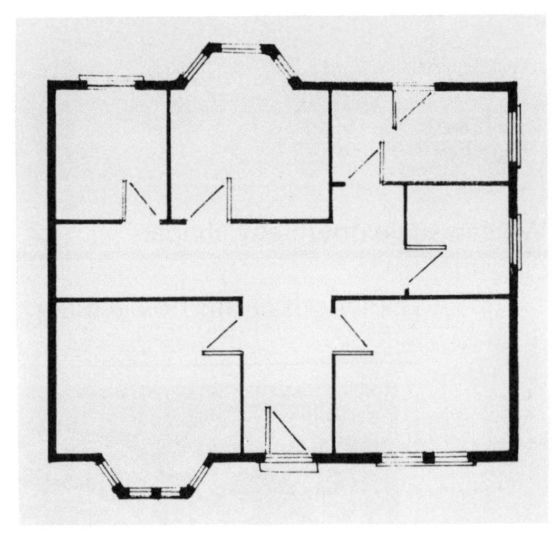

ASSIGNMENT 7 Workbook page 49

You should now be able to draw a rough plan of your own home.
First you will need to list the different rooms.
Remember to show all the doors and windows!

ASSIGNMENT 8

Workbook page 50

Scale Drawings

You will realise that the plans you have drawn have all been much smaller than true-life size.

This is a true-life size drawing of an egg cup.

This egg cup is drawn four times smaller, or $\frac{1}{4}$ of the size of the real egg cup. We say it is **drawn to scale**.

Instead of writing '$\frac{1}{4}$ of the original size', we can write 'the scale is 1:4'.
1:4 is read 'one to four'.

1:3 (one to three) means $\frac{1}{3}$ of the original size.
1:6 (one to six) means $\frac{1}{6}$ of the original size.
1:10 (one to ten) means $\frac{1}{10}$ of the original size.

ASSIGNMENT 9

Workbook page 50

We can scale down any shape.

The length of this box is 6 cm.

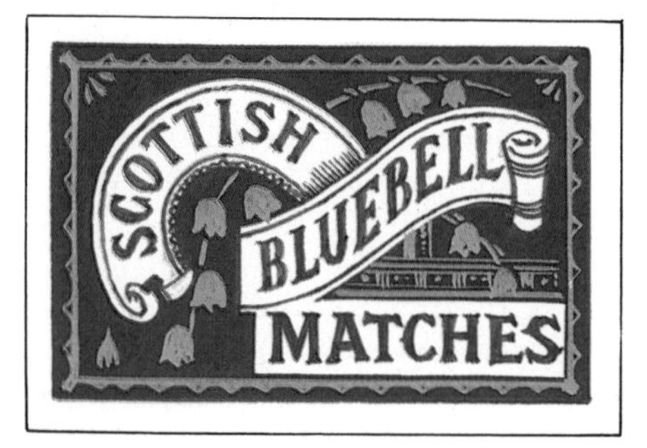

The length of this is 2 cm.

It is $\frac{1}{3}$ of the original size, so the scale is 1:3.

This shape is 5 cm long.

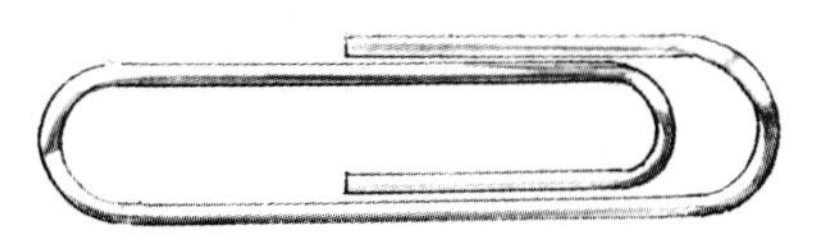

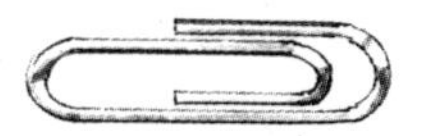

This is 2.5 cm long.
It is $\frac{1}{2}$ the size, so the scale is 1:2.

ASSIGNMENT 10

Workbook page 51

When you make a scale drawing of an object, you write the scale beside it.
On an architect's plan you will also find a scale.
Look at this plan of a bungalow. The plan is $\frac{1}{100}$ actual size.

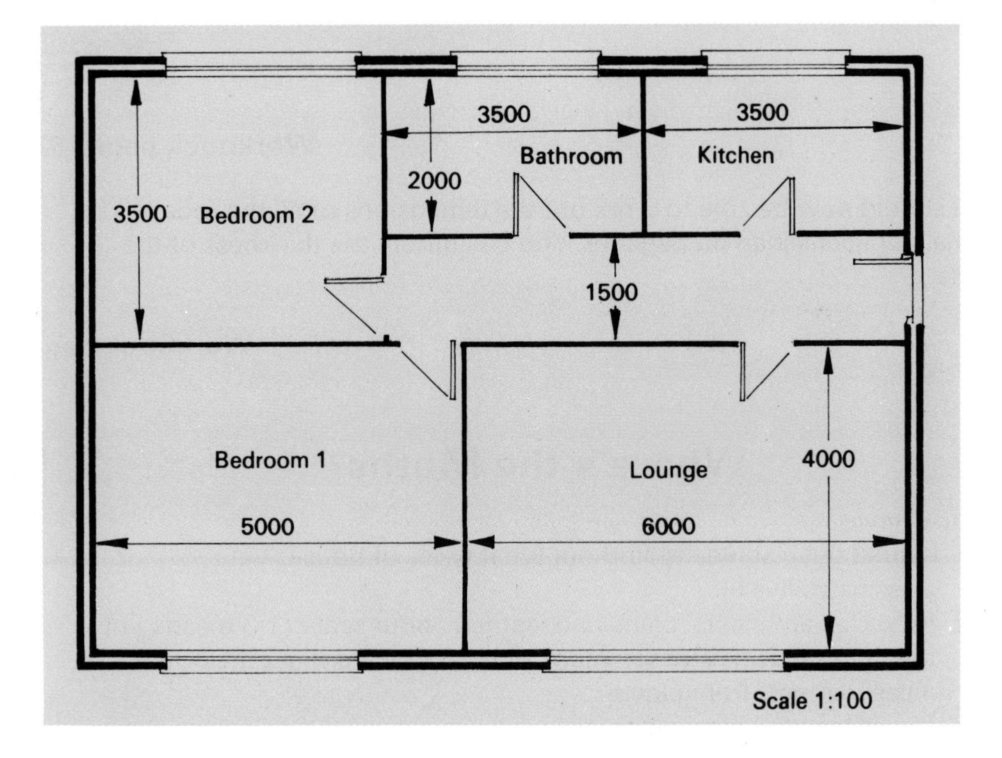

As well as writing the scale on the plan, the architect marks in the actual measurements or dimensions of the rooms. The dimensions on the plan are all in millimetres.
Look carefully at the architect's plan again. You will see that not all the measurements are given. However, you should be able to work them out from the plan.

Look at the rectangle $ABCD$.
Line $AB = 6$ cm
Line $BC = 4$ cm
We know that the opposite sides of a rectangle are equal, so line $DC = 6$ cm and line $AD = 4$ cm.

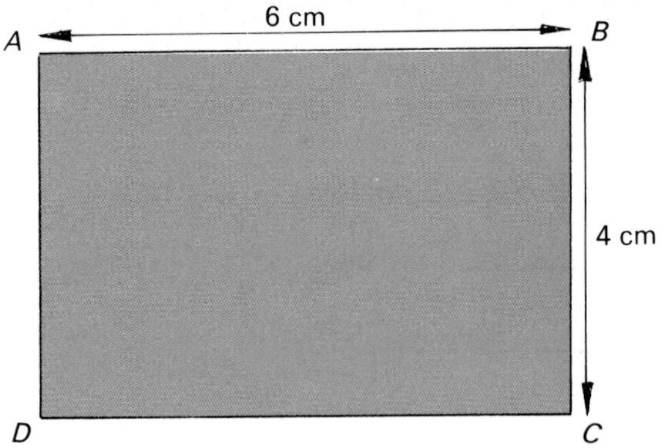

A 6 cm B
4 cm
D C

In this shape we know that line $AB = EC$, so $EC = 7$ cm.
The space $DC = 7 \text{ cm} - 5 \text{ cm}$
$= 2 \text{ cm}$

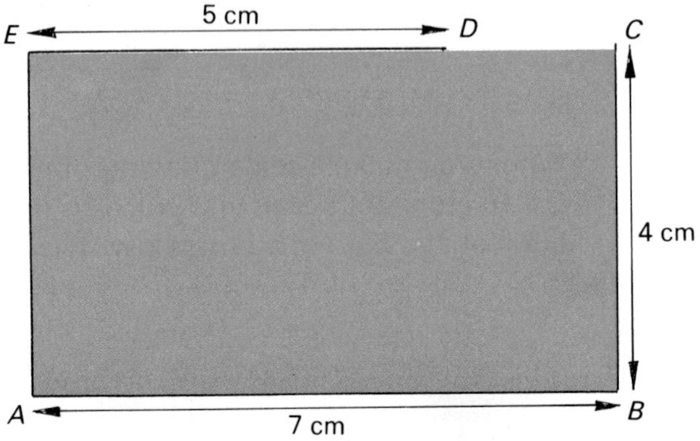

E 5 cm D C
4 cm
A 7 cm B

ASSIGNMENT 11

Workbook pages 52–53

You should now be able to work out the dimensions of all the rooms in the architect's plan on page 49. You can ignore the thickness of the walls.

ASSIGNMENT 12

Workbook page 53

Where's the Maths?

You have:

1 carried out a survey to find out what types of houses your classmates live in.
2 looked at architects' plans and learned about scale (1:3 means $\frac{1}{3}$ of the original size), measurement, and how to work out unknown measurements from plans.
3 drawn simple house plans.

Words to Remember

architect **plan** **scale**

If you were going to the USA, how would you travel?
You could fly Concorde and be there in $3\frac{1}{2}$ hours . . .

or you could take a more leisurely trip in the *QE2.*

Most of our everyday travelling is done by car, by bus or on foot. But there are many other ways of travelling, or **modes of transport** as they are usually called. Here are pictures of some.

How many of these have you used?

ASSIGNMENT 1

Workbook page 55

A Statistical Survey

Find out how the pupils in your class come to school.

ASSIGNMENT 2

Workbook page 56

We now want to illustrate the information you have collected. You have already drawn bar graphs and block graphs. This time you are going to draw a **pictograph**.
Look at the data in the table below.
Then look at the pictograph which illustrates this data.

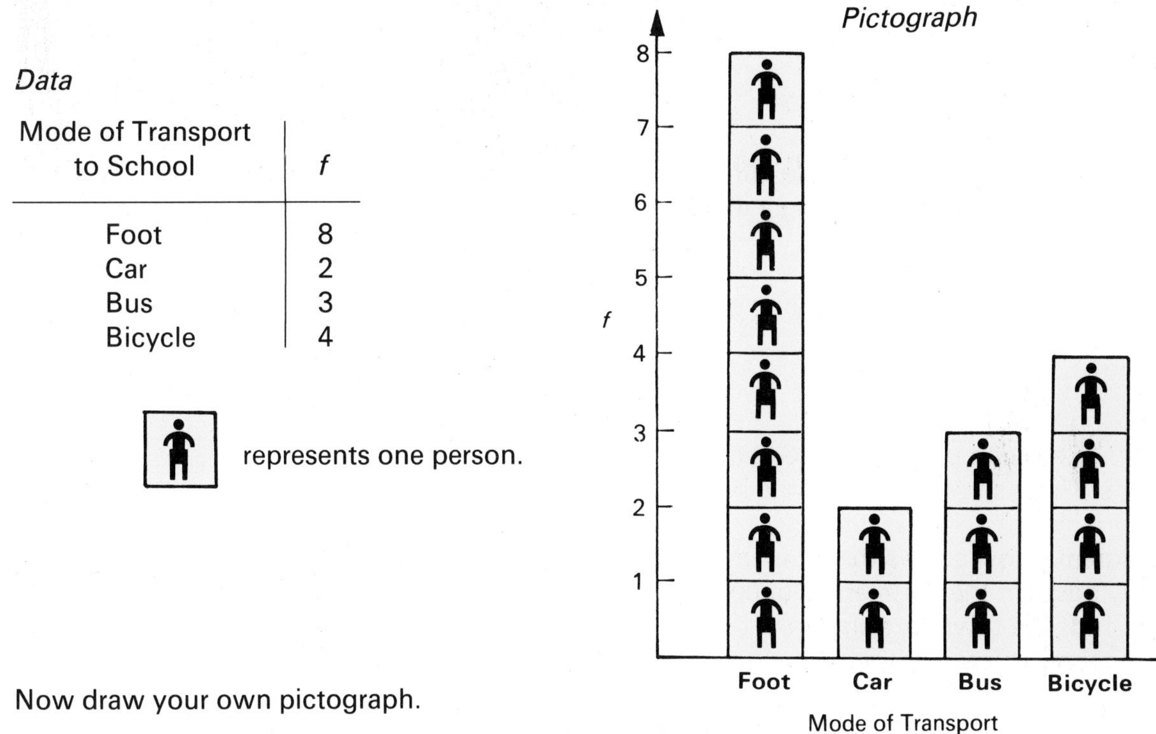

Data

Mode of Transport to School	*f*
Foot	8
Car	2
Bus	3
Bicycle	4

[pictogram] represents one person.

Now draw your own pictograph.

ASSIGNMENT 3

Workbook page 56

Going on a Journey

If you are going on a journey, you should know:
1 how you are travelling (the mode of transport);
2 where you are going (the direction);
3 how far you are travelling (the distance).
You have already looked at modes of transport.
Now let's learn something about direction.

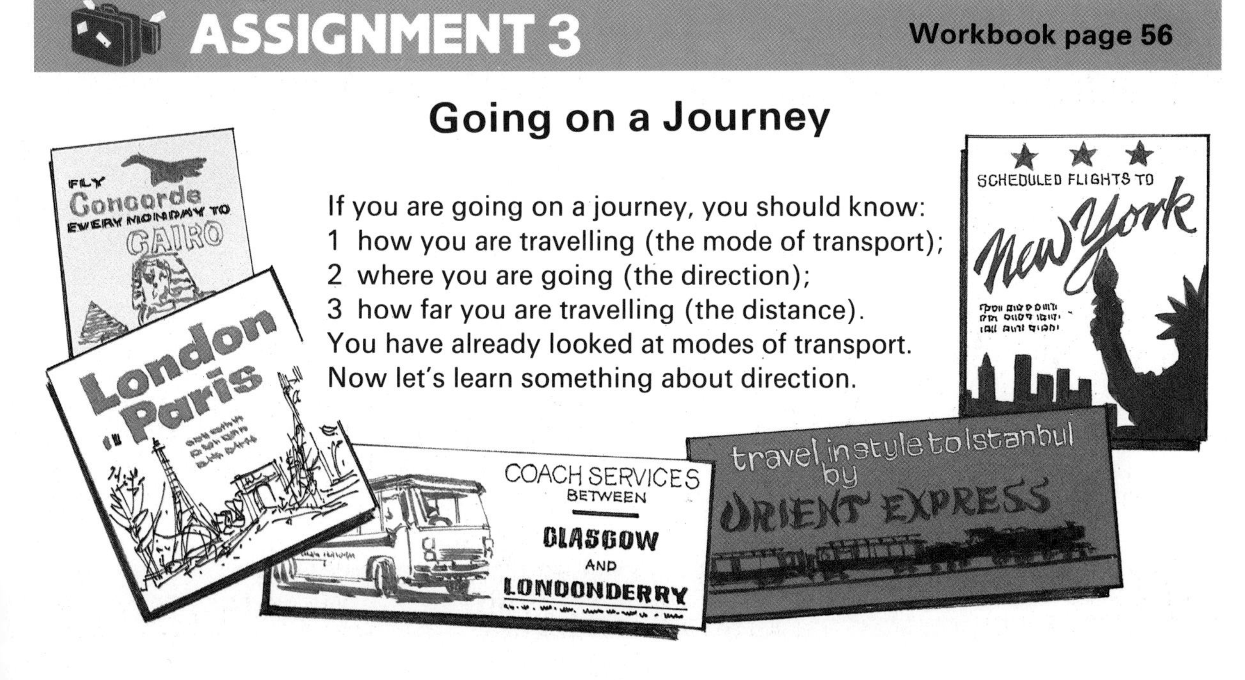

Direction

You often go:
1 from home to school;
2 from school to the playing fields;
3 from the disco to your home.
You will also have gone further afield, perhaps even abroad on a holiday.

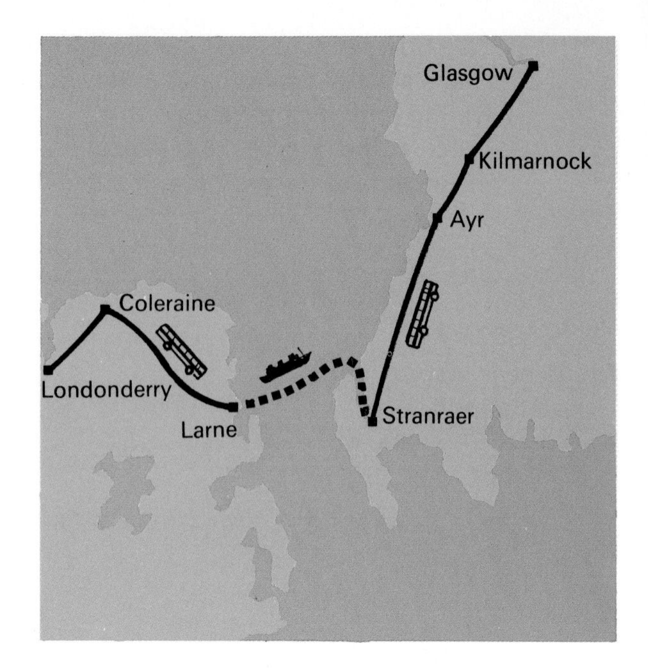

We can draw diagrams to illustrate different journeys. Here is an example using a map.

Sometimes bus companies produce simplified diagrams. These show the routes the buses take.
The diagram below shows the different routes taken by buses in the Perth and Dundee area in Scotland. The towns are shown in the correct geographical position.

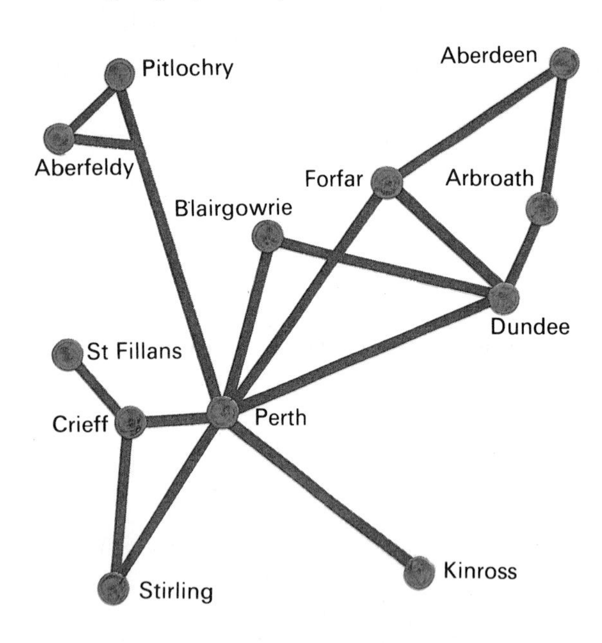

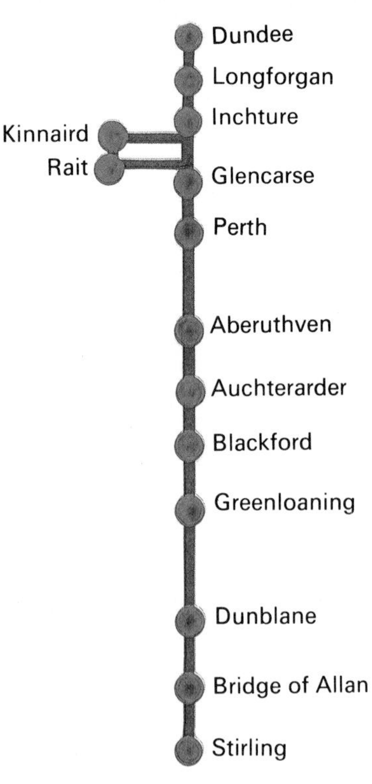

The diagram above shows the bus stops between Dundee and Stirling. It is drawn in a straight line to make it simple to follow.

ASSIGNMENT 4 — **Workbook page 56**

These types of diagrams are called **route networks**. A route network is a diagram which shows routes or roads between towns.
Here is a route network showing the main routes in a part of South Wales. Compare it with the map on the left.

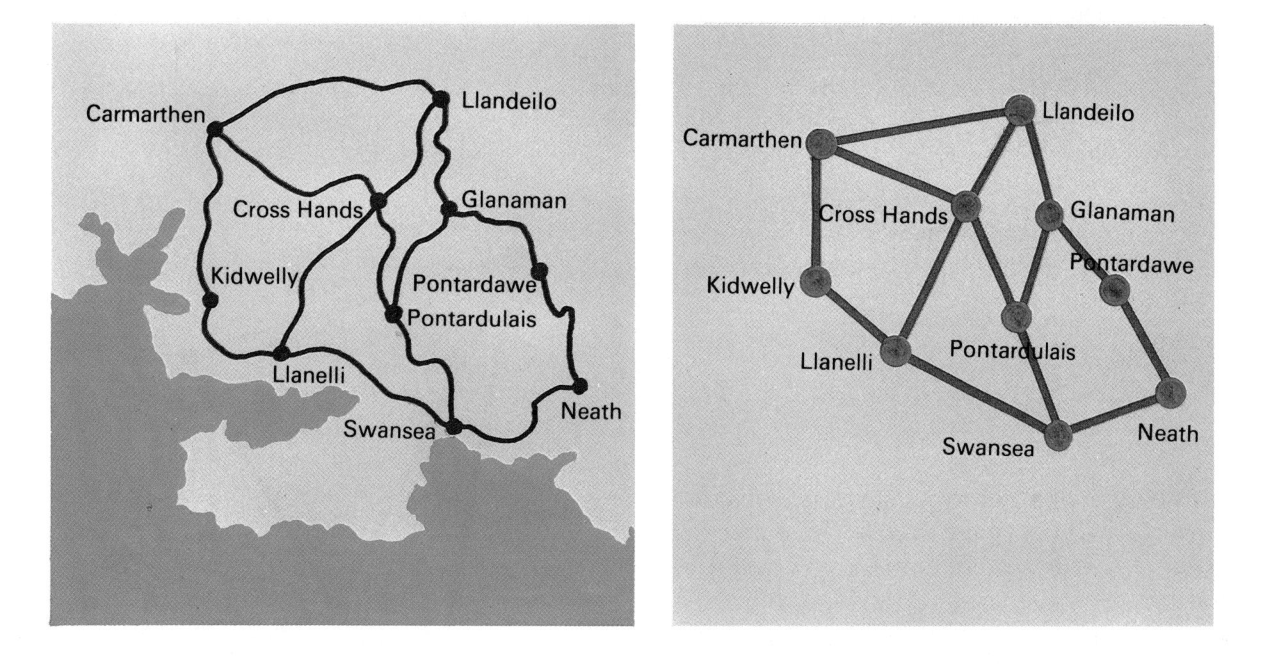

ASSIGNMENT 5

Workbook pages 57–58

Distance or Length of a Journey

We have looked at length before. We have measured:

1 the length and breadth of a rectangle;

2 the radius of a circle.

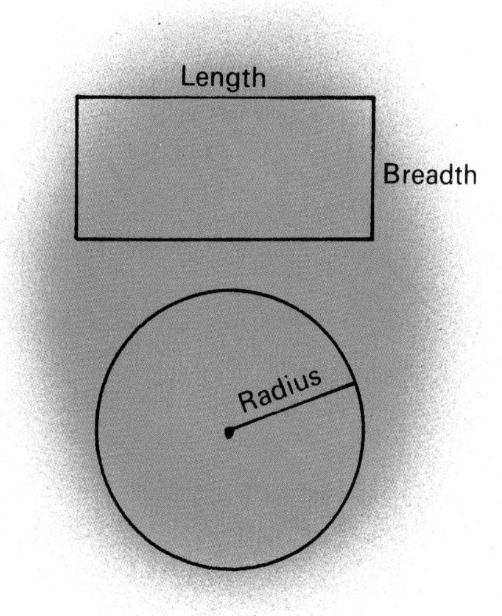

Remember you even made your own measuring tape.

The units we used were millimetres, centimetres and metres.
We would measure the length of an insect in millimetres.
We would measure the length of a pencil in centimetres.
We would measure the length of a football pitch in metres.
But we need a much larger unit to measure the distances between towns.
The unit we use is the **kilometre**. The abbreviation for kilometre is **km**.

1000 metres = 1 kilometre	or	1000 m = 1 km

Ask your teacher how far one kilometre is from your school. This will give you an approximate idea of the distance.

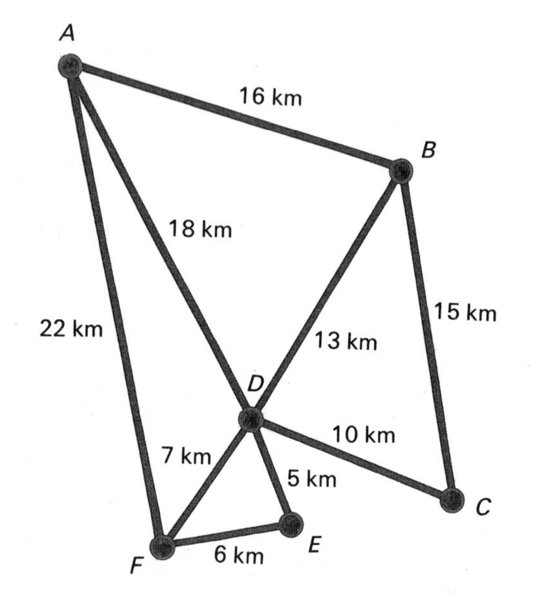

Let's look at another route network. This one shows a network of six towns. The distances between the towns are written next to the lines. The distance from *B* to *C* is 15 km.
The distance from *B* to *F* through *D* is
13 km + 7 km = 20 km.

ASSIGNMENT 6 — Workbook pages 58–59

If we write down the journey from *A* to *B*, like this: $A \rightarrow B$
it will save us some time.
$B \rightarrow C$ means a journey from *B* to *C*.
$D \rightarrow E$ means a journey from *D* to *E*.
$F \rightarrow D \rightarrow B$ means a journey from *F* to *B* going through *D*.
$A \rightarrow B$ is 16 km. $C \rightarrow D$ is 10 km. $A \rightarrow D \rightarrow B$ is 18 km + 13 km = 31 km.

ASSIGNMENT 7 — Workbook pages 59–60

Other Types of Route Networks

The distance between towns is sometimes shown like this:

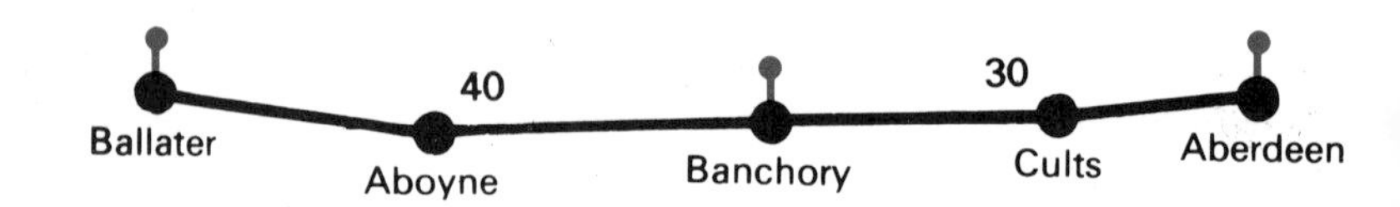

Only the distance between the main towns is given. It would be too complicated to put in the distances between all the towns.

This sign tells us where to start and finish:

From the example shown, Ballater→Banchory is 40 km,
Banchory→Aberdeen is 30 km.

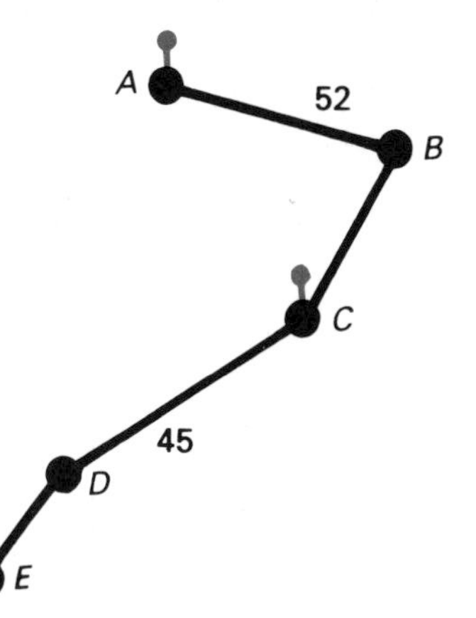

Here is another example.

A→*B*→*C* is 52 km.
C→*D*→*E* is 45 km.

ASSIGNMENT 8

Workbook pages 60–61

Distance Tables

In books, atlases and road maps you often find tables of distances between towns. This is a route network. The distances are given in kilometres.

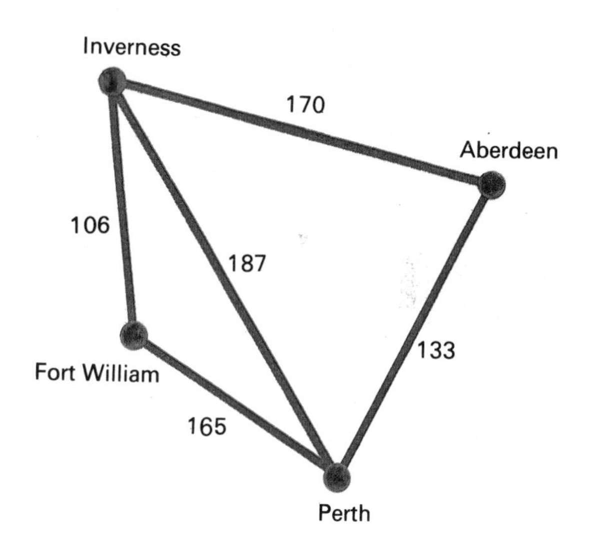

The information on distances can be made into a table.

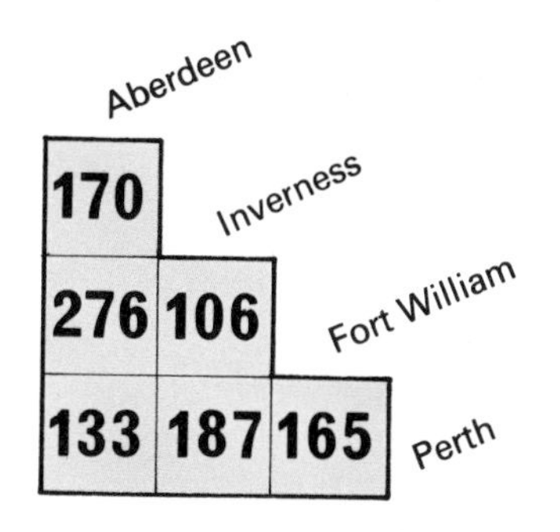

Aberdeen			
170	Inverness		
276	106	Fort William	
133	187	165	Perth

To find the distance between Aberdeen and Perth, run one finger down the 'Aberdeen line' and another along the 'Perth line' until they meet.
Aberdeen→Perth is 133 km.

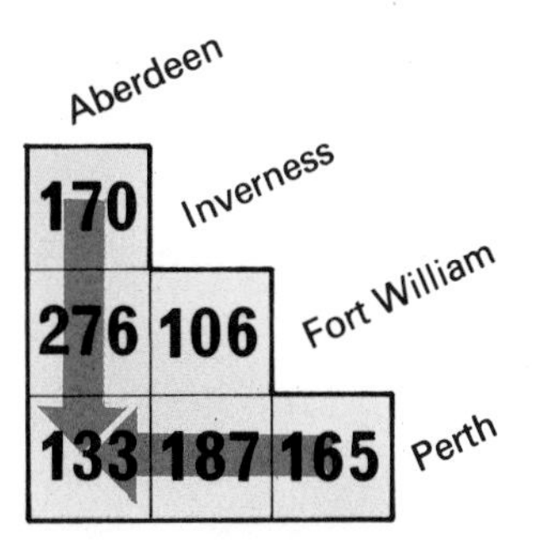

Aberdeen			
170	Inverness		
276	106	Fort William	
133	187	165	Perth

Here is a distance table (in kilometres) between four towns.
$A \to B$ is 3 km.
$A \to C$ is 4 km.
$B \to D$ is 7 km.
$A \to D$ is 6 km.

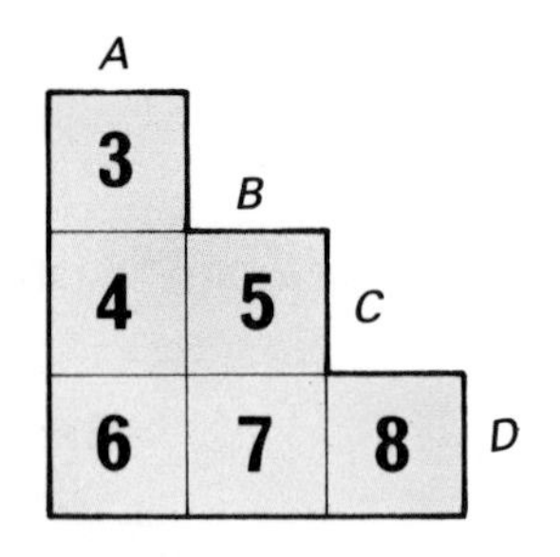

ASSIGNMENT 9

Workbook pages 61–62

Where's the Maths?

You have:

1 carried out a survey and collected statistics.
2 illustrated the data in a pictograph.
3 analysed the data.
4 looked at the distance and direction components of a journey.
5 learned about different types of diagrams illustrating journeys.
6 drawn route networks.
7 taken information from tables of distances.

Words to Remember

pictograph **distance** **direction**

route network **kilometre (km)**

Measurements to Remember

10 mm = 1 cm **100 cm = 1 m** **1000 m = 1 km**

7. MONEY

What would you do if you won a lot of money?
What would you spend the money on?
Do you think it would change your life . . .
for the better?

The diagram shows some of the things a family needs money for.

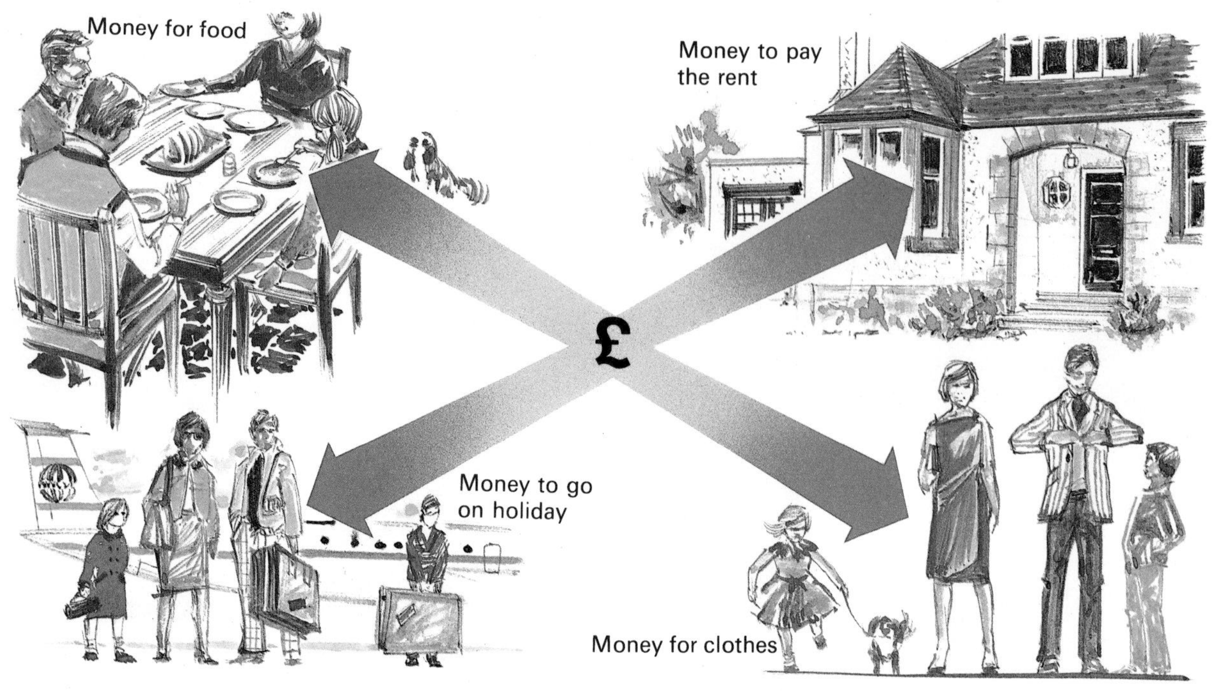

There are lots more ways of spending money!

ASSIGNMENT 1

Workbook page 64

Let's find out about the actual money we use. Coins have been used as money for a very long time. The Greeks were probably the first people to make coins. That was around 700 BC.

This is a very old Greek coin made of silver. It is called a *dekadrachm* and was made about 400 BC.
Both sides of the coin are shown. The 'heads' side of a coin usually shows the head of the ruler. Many different designs are used on the 'tails' side.

Roman coins can be seen in many museums. Here are both sides of two coins from about 2000 years ago. They were called *denarii*.

The emperor on this one is Augustus.

The emperor on this one is Nero.

Before coins, other objects such as shells or bright stones were used. Today in some countries they still use shells!

Even further back in time people exchanged objects. They did not pay money for them. This was called **barter**.

Today we use coins and notes. The unit of money in the United Kingdom is the **pound**. One pound is divided into 100 **pence**. We say that our **currency** is pounds and pence. The abbreviation for pound is **£**. The abbreviation for pence is **p**.

£1 = 100p

Here is one of the £1 coins we use today. Both sides are shown. The £1 coin was introduced in 1983. Can you think why it was needed instead of a note?

Long ago coins were made of silver and gold. Nowadays the metals used are copper, nickel and zinc. These are much cheaper metals. Here are the coins we use in Britain today. They are shown actual size.

There are seven altogether. Four are silver coloured and two are bronze coloured. The £1 coin is a golden colour. Each coin has an impression of the Queen on the 'heads' side. The 'tails' sides shown here have various emblems. Do you know what these are?

ASSIGNMENT 2

Workbook pages 64–65

Here are some sets of coins.

These coins together add up to 70 pence.
We write this as 70p.

This set adds up to 112 pence. We could write 112p but usually we write £1.12.

This set adds up to one pound and 35 pence.
We write this £1.35.

ASSIGNMENT 3

Workbook page 66

If a sum of money is less than one pound, we usually write 5p or 23p or 64p, etc. If the money is more than one pound, we usually write £1.21 or £5.60 or £6.40, etc.

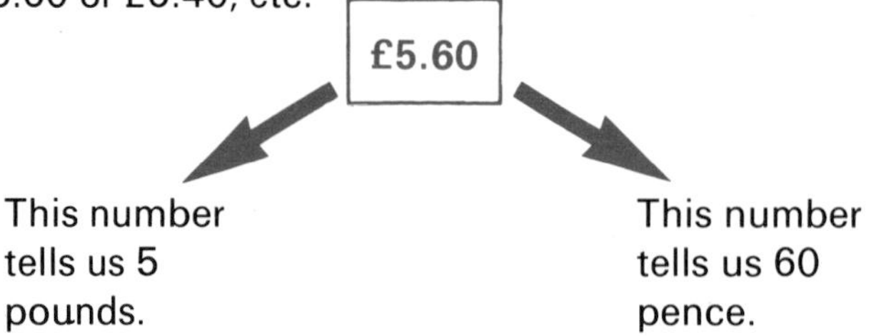

We never use both the £ sign and the p sign with the same amount of money.

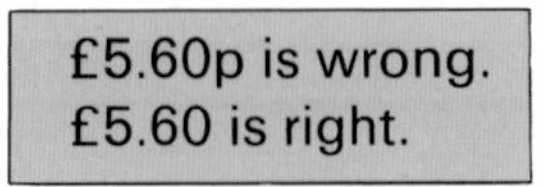
£5.60p is wrong.
£5.60 is right.

Remember this!

Here are some examples to show you how to change pence to pounds.

120p = 100p + 20p	356p = 300p + 56p	70p = £0 + 70p
= £1 + 20p	= £3 + 56p	= £0.70
= £1.20	= £3.56	

ASSIGNMENT 4

Workbook page 67

Notes

Notes are used for larger amounts of money. Here is a £5 note. Look at the details on it.

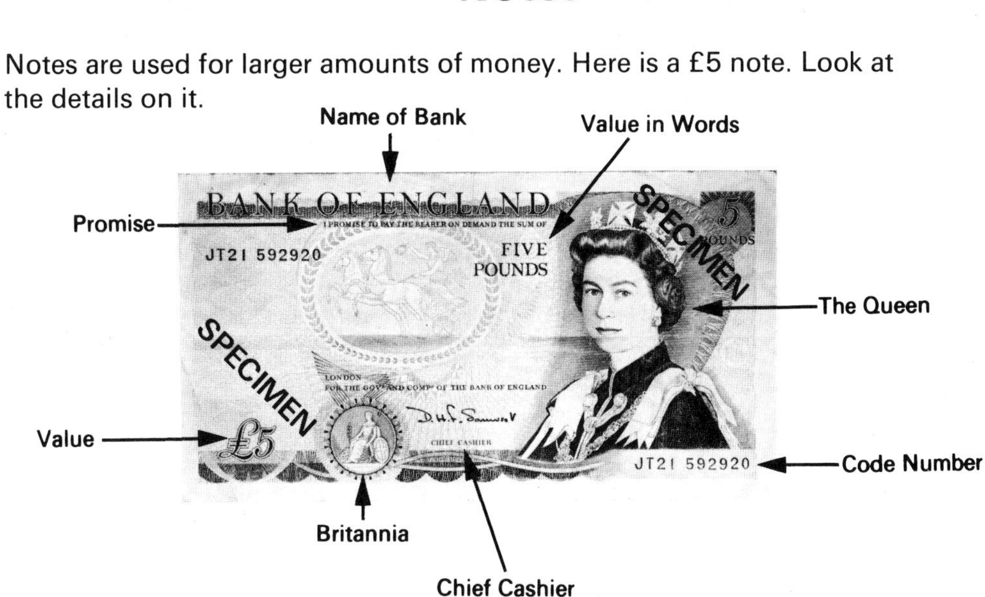

All these fine details and the design make the note very difficult to copy. Some people try! They are called counterfeiters.
You can also get £10, £20, £50 and even £100 notes. Different banks produce different notes, but the details on the notes are very similar.

ASSIGNMENT 5

Workbook page 68

Change

When you buy something in a shop, you don't always have the exact money.

If you gave the shop assistant £13 for these jeans, you would get 1p change.

If you gave the shop assistant £5 for this lipstick, you would get £2.35 change.

At checkouts in certain shops, your change is displayed on the till.

Some shop assistants place the change into your hand all at once. Most shop assistants return the change step by step.

Jim decides to buy a shirt costing £8.53. He gives Kate, the assistant, a £10 note. She counts the change into Jim's hand, saying 'and two makes 8.55, 8.60, 8.80, 9, and one makes £10.'
Can you work out what coins she has given him and what his total change is?

ASSIGNMENT 6

Workbook page **68**

Spending Your Money

How much pocket money are you given each week? Do you think it's enough? Perhaps you earn some extra money by delivering milk or newspapers, or working in a shop.
How do you spend your money? Here are four ways you might spend it.

Cinema

Space-invader Machines

Your Cash

Tickets for a Football Match

Personal Stereo

But there are lots of ways to spend money!

ASSIGNMENT 7

Workbook page 69

Let's look at one way your money is spent – clothes. Fashions change all the time. Just look at the contrast in these fashions!

Let's look at how much one of these outfits cost at the time.

Hat 4/6
Blouse 1/6
Skirt 2/-

Of course, these prices are in 'old' money. In Britain, the decimal system of money was introduced in 1971.

Here are the prices again – this time in decimal money.

Hat $22\frac{1}{2}$p
Blouse $7\frac{1}{2}$p
Skirt 10p

What do you think clothes like these would cost today?

Let's look at some more modern clothes and what they cost.
This is a casual outfit for a girl or boy!

T-shirt £3.50
Jeans £15.00
Training shoes £12.75

If we wanted to find out the total cost of this outfit,
we would have to add all these costs together.
The easiest way to do this is to set it out as a sum.

Remember, when you are writing down sums of money,
to always keep the decimal points under one another.

	£3.50
	£15.00
	+ £12.75
Total	£31.25

ASSIGNMENT 8

Workbook pages 69–70

Where's the Maths?

You have:

1 looked at how money is spent.
2 learned something about the history of money.
3 looked at British coins and notes.
4 revised your units of currency (100p = £1).
5 worked out the value of sets of coins.
6 worked out change due and learned how to give it step by step.
7 looked at how money is spent on clothes.
8 practised adding up sums of money.

Words to Remember

pound pence barter currency

8. FOOD

Our eating habits have changed quite a lot over the years. Look at these two scenes.

Let's look at some meals offered in a restaurant today.
Most people will have seen a **menu**.
A menu lists the selection of food you can choose from.

Here is a typical menu.

The menu is divided into different types of food.

You could say the different types of food are **classified** into three courses.

What would you choose from this menu?

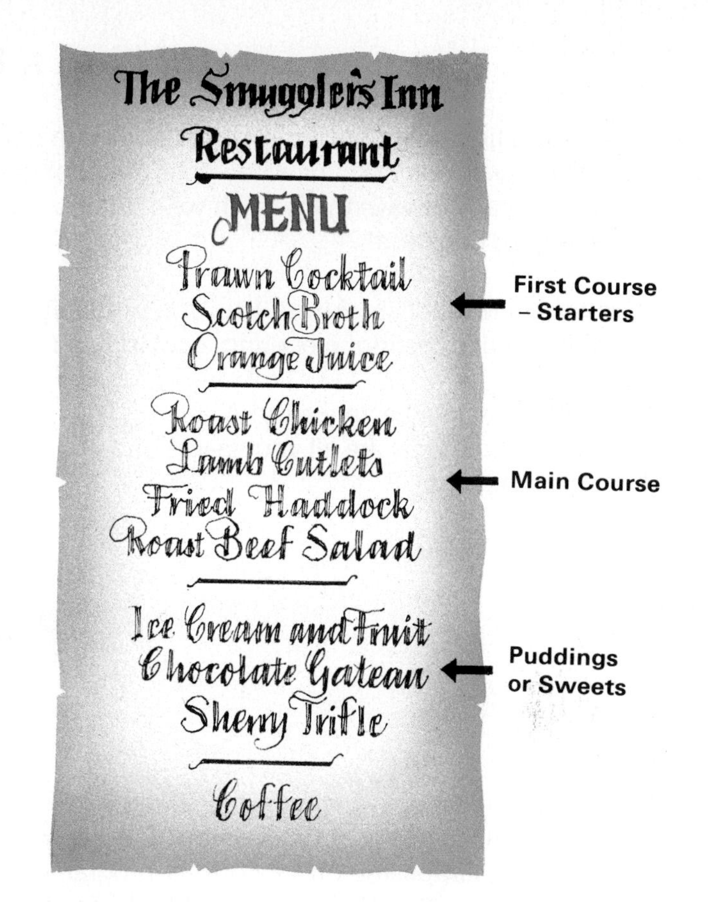

ASSIGNMENT 1

Workbook page 72

Now try making up your own menus with the food *you* like!

ASSIGNMENT 2

Workbook page 73

A Statistical Survey

Find out the most popular meals of those in your class.
First choose your own six favourite starters, main courses and sweets.
Then collect the class data and illustrate it in a pictograph.

ASSIGNMENT 3

Workbook pages 73–74

Cooking a Meal

Most of us enjoy going out for a meal but we can't afford to do it too often. We usually eat at home. Who does the cooking in your house? It used to be nearly always the women who did the cooking. Nowadays most men can cook too!

Have you done much cooking? If so, you probably already know something about costing, preparing, cooking and serving a meal.

Cooking a meal can be very easy! A can opener and some pots are all you need – and of course some tins!

Let's see what maths is involved in cooking a meal.
Perhaps you have heard of Delia Smith. She has written a few cookbooks and appeared on TV cookery programmes. At the front of each of Delia's cookery books you will find conversion tables like these:

Oven Temperatures		
Mark 1	275°F	140°C
2	300	150
3	325	170
4	350	180
5	375	190
6	400	200
7	425	220
8	450	230
9	475	240

Volume	
2 fl oz	55 ml
3	75
5 ($\frac{1}{4}$ pt)	150
10 ($\frac{1}{2}$ pt)	275
15 ($\frac{3}{4}$ pt)	425
20 (1 pt)	570
$1\frac{3}{4}$ pt	1 litre

Weights	
$\frac{1}{2}$ oz	10 g
1	25
$1\frac{1}{2}$	40
2	50
$2\frac{1}{2}$	60
3	75
4	110
$4\frac{1}{2}$	125
5	150
6	175
7	200
8	225
9	250
10	275
12	350
1 lb	450
$1\frac{1}{2}$ lb	700
2 lb	900
3 lb	1 kg 350 g

They compare the old non-metric measures and the newer metric measures. Before you try to understand how to use the tables, you should know what is meant by volume, weight and temperature. Let's start with volume.

Volume

Here is a list of ingredients for leek and potato soup.

4 large leeks
2 medium-sized potatoes
1 medium-sized onion
850 ml chicken stock
300 ml milk
Knob of butter
Seasoning

In this list there are liquids and solids. The experienced cook should know the sizes of the solids – large leeks and medium-sized potatoes. The liquids (850 ml chicken stock and 300 ml milk) must be measured.

To measure the liquids we use a measuring jug. Look at this measuring jug. On the jug there is a **scale**. The scale tells us how much space the liquid takes up. We are measuring the **volume** of the liquid.

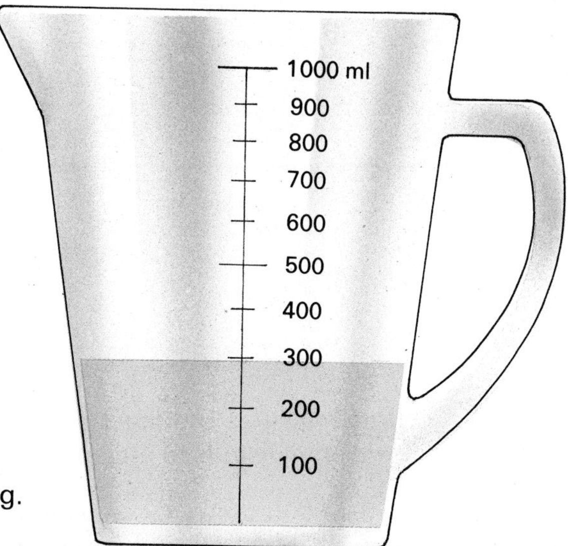

The units we use are litres and millilitres.

1000 millilitres = 1 litre

The abbreviation for millilitre is **ml**.
The abbreviation for litre is *l*.

So

1000 ml = 1 *l*

There are 300 ml of liquid in the jug.

To help you understand how much liquid there is in 1 litre, let's use a familiar container. A small yogurt carton usually holds 150 ml, so seven measures would make just over 1 litre.

If you buy any type of liquid, the container should have the volume printed on it. Remember, the volume is the amount of space the liquid takes up. Here are some examples.

Both millilitres and litres are used. Make sure you know how to convert millilitres to litres and back again!

ASSIGNMENT 4

Workbook page 75

Here are some measuring jugs with various scales.
How much liquid is there in each of them?

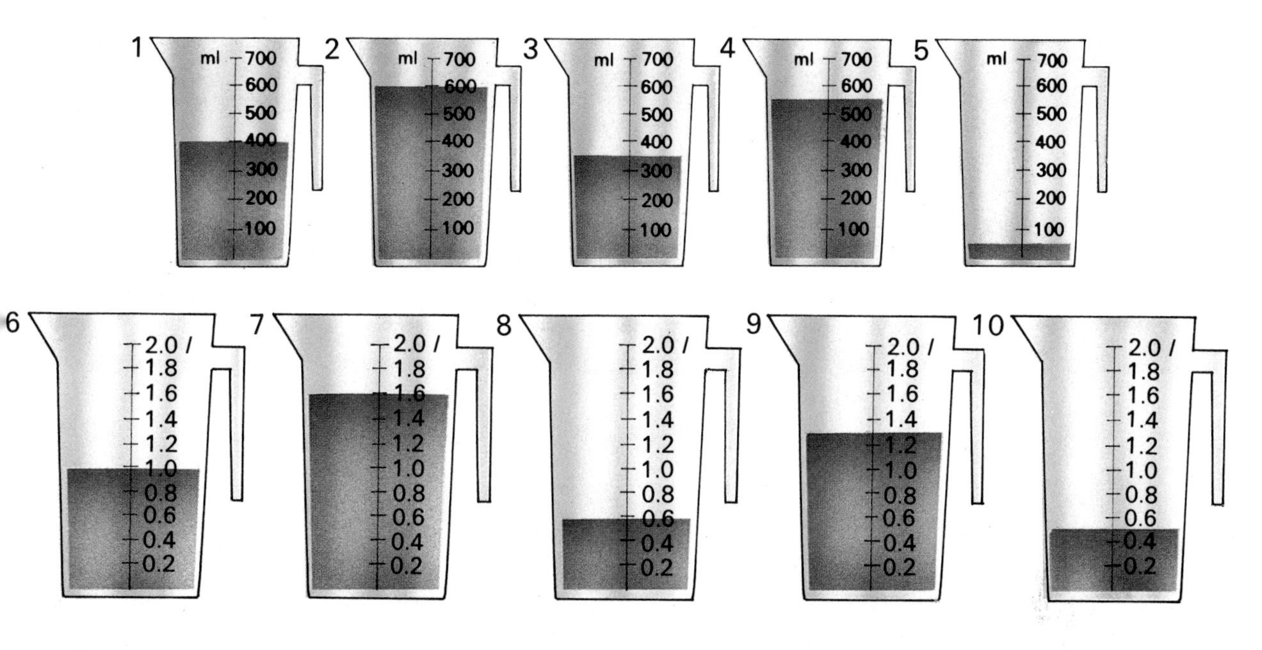

ASSIGNMENT 5

Workbook page 75

Look at these bottles. Which one do you think would hold the most?
We can use a measuring jug to find the volume of each. Here's what to do.

1 Fill the bottle with liquid.

2 Pour the water from the bottle into the jug.

3 Read off the measurement on the scale of the jug.

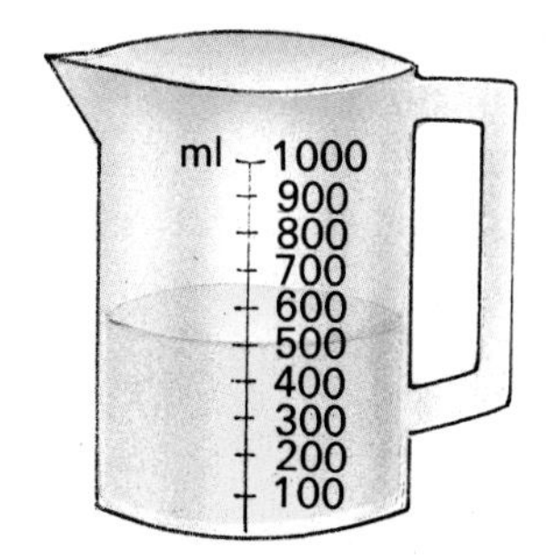

ASSIGNMENT 6

Workbook page 76

You should now have a better idea about how much one litre of liquid is. How good are you at guessing how much any container will hold?

ASSIGNMENT 7

Workbook page 76

Weight

We've looked at the ingredients for a starter – leek and potato soup. Now let's see what we're going to have for the main course! Here is a list of ingredients for steak and kidney pudding.

250 g self-raising flour
Pinch of salt
125 g suet
Water to mix
750 g stewing steak
150 g kidney
25 g flour
Seasoning
250 ml beef stock

We know how to measure the liquid in this recipe. Measuring a quantity of something solid is different from measuring a liquid. We measure the **weight** of the solids, e.g. the flour, suet, steak and kidney. The units we use are **grams** and **kilograms**.

1000 grams = 1 kilogram

The abbreviation for gram is **g**. The abbreviation for kilogram is **kg**.

So

1000 g = 1 kg

To find out how heavy an object is, we use a weighing machine. There are many different types of weighing machine. Here are two types that actually use weights on the 'pans'.

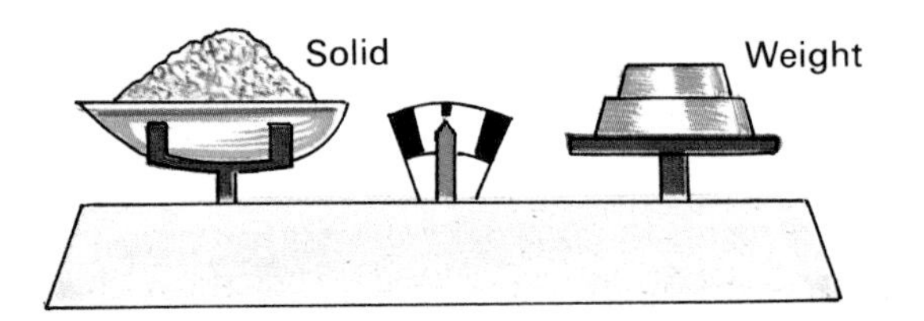

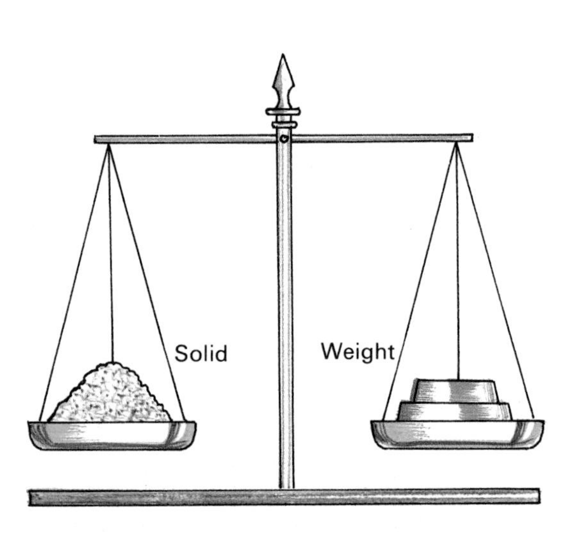

Nowadays, most kitchen scales have one pan with a dial to show the weight. They also come in different shapes and sizes.

Most food packages have the weight of the contents printed on the outside. Here are some examples. Both grams and kilograms are used. Make sure you know how to convert grams to kilograms and back again!

ASSIGNMENT 8

Workbook page 77

Here is the scale on a weighing machine.
This weighing machine weighs up to 2.5 kg.
The large numbers are kilograms.
The small numbers are grams.
Each of the small divisions is 20 g.
The pointer tells us the weight of the object.
The object here weighs 800 g.

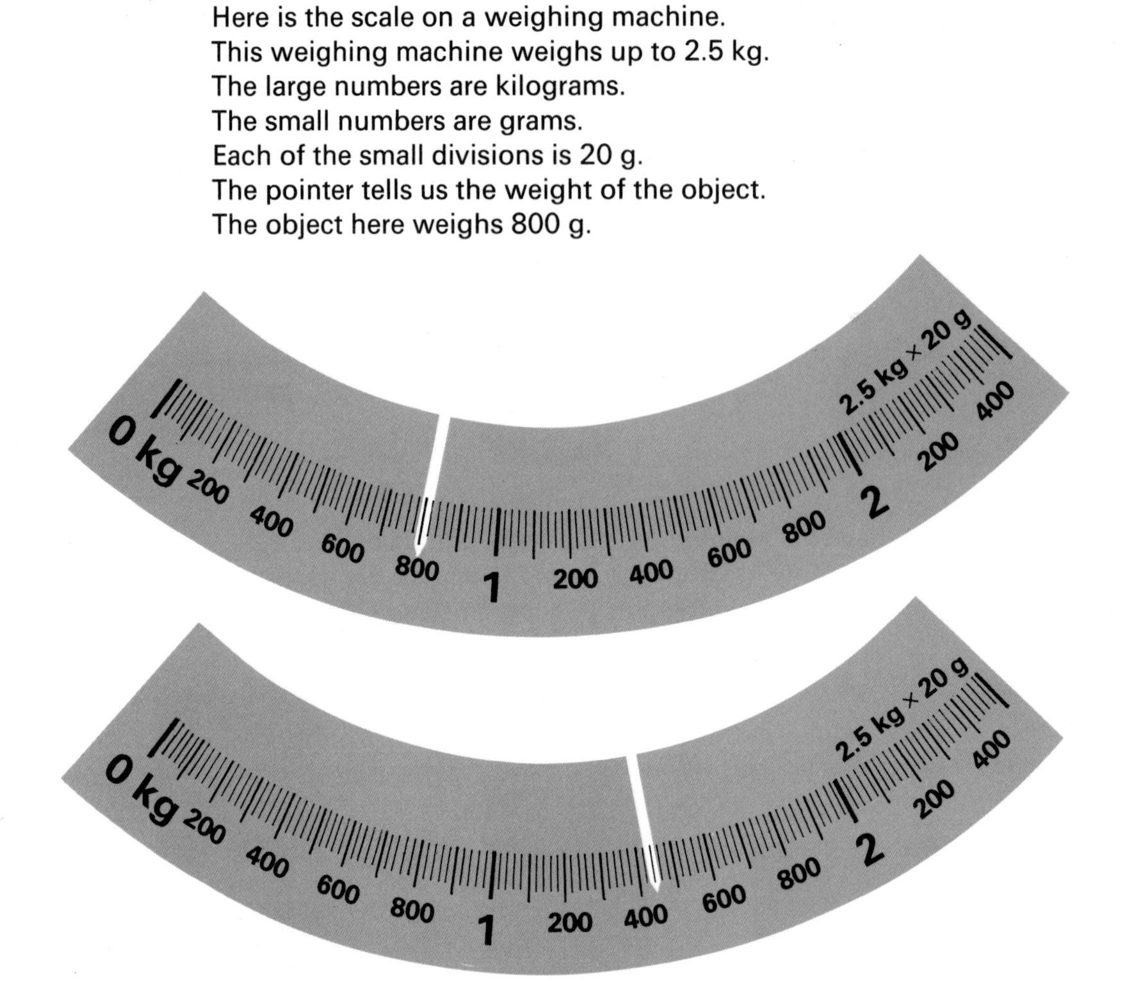

The object here weighs 1440 g or 1.44 kg.

ASSIGNMENT 9

Workbook pages 77–78

Here is a different scale from another weighing machine. This time each of the small divisions is 50 g. The weight here is 1500 g or 1.5 kg.

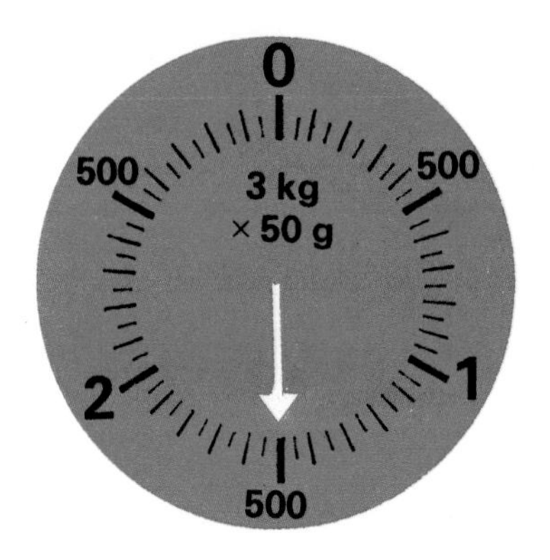

ASSIGNMENT 10

Workbook page 79

There are many other types of kitchen scales. Find out what kind you have at home. Now that you've had some practice at reading scales, try weighing some real objects!

ASSIGNMENT 11

Workbook page 80

You should now have a good idea of how heavy one kilogram is. Find out how good you are at guessing weights.

ASSIGNMENT 12

Workbook page 80

By law all foods sold in a tin or packet should have the weight printed on the label. To finish off your work on weight, see how many different labels you can collect.

ASSIGNMENT 13

Workbook page 80

Where's the Maths?

You have:

1. classified data.
2. drawn a pictograph.
3. measured the volume of a liquid.
4. used a measuring jug and read the scale in millilitres and litres.
5. changed millilitres to litres and litres to millilitres.
6. guessed the volume of a container.
7. measured the weight of a solid.
8. used a weighing machine and read the scale in grams and kilograms.
9. changed grams to kilograms and kilograms to grams.
10. guessed the weight of a solid.

Words to Remember

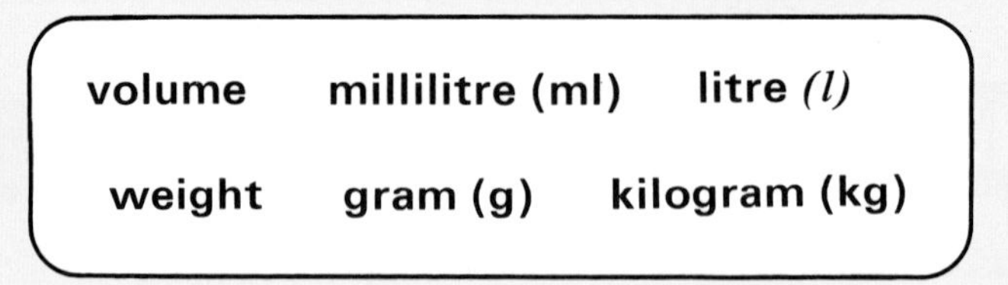

Measurements to Remember

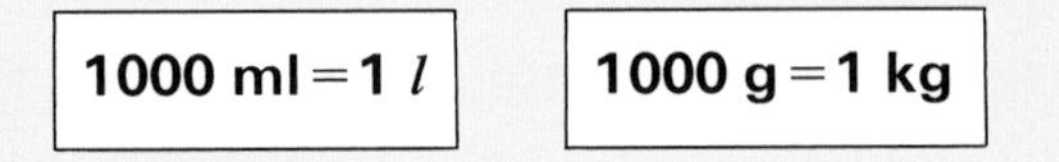

Acknowledgements

The author and publishers would like to thank the following for permission to reproduce the photographs on the pages listed below.

Page 5 The BBC
Page 6 Central Independent Television plc
Page 13 Sporting Pictures (UK) Ltd
Page 21 *(top)* The Scottish National Orchestra
Page 21 *(bottom)* London Features International Ltd
Page 23 *(bottom right)* The Science Museum (British Crown Copyright)
Page 23 *(bottom left)* Bang & Olufsen (UK) Ltd
Page 29 Space Frontiers Ltd/NASA
Page 38 *(top and bottom left)* BBC Hulton Picture Library
Page 38 *(bottom right)* Space Frontiers Ltd/NASA
Page 44 *(top)* The National Trust for Scotland
Page 51 *(top)* Camera Press Ltd
Page 51 *(bottom)* Cunard Line
Page 60 The British Museum

Thanks are also due to the Royal Mint and the Bank of England for permission to reproduce the British coins and five pound note on pages 61–63.